Research Project Management:

The Research Proposal

Bryan McKersie

Created by Kindle Direct Publishing

Edition 1.0, Nov. 2019
Copyright © 2019 by Bryan McKersie.
All Rights Reserved.

ISBN 9781719402200

To Marie

For her patience and support

Table of Contents

Introduction 7

PART One - Research Projects 13
    Expectations of Research Projects 14
    Cycle of Innovation 17
    Key Points 20

PART Two - Research Project Management 21
    The Workplan 22
    Project Management 25
    Research Projects are Different 32
    The Ten Principles of Research Project Management 42
    The Research Proposal 48
    Key Points 51

PART Three - Creating Ideas 53
    Creative Thinking 54
    Free Association Techniques 57
    Structured Association Techniques 58
    Combination Techniques 60
    Analogy 60
    Feedback 61
    Convergent Thinking 63
    Selection Criteria 65
    Review Panels 68
    Group Decision-making 71
    Key Points 75

PART Four - Finding Solutions to Problems 77
    The Problem 78
    Observation Projects 83
    The Vision 86
    Literature Reviews 90
    The Scientific Model 94
    Modeling Projects 103
    Key Points 106

PART Five - Drafting a Research Proposal 109

The Objective	110
The Research Concept	115
Understanding Sponsors	123
Understanding Customers	129
Selling the Research Proposal	132
Key Points	135

PART Six - The Scientist's Toolbox 137

Overview	138
Tool # 1 Brainstorming	140
Tool # 1.1 Face-to-Face	142
Tool # 1.2 Cards	142
Tool # 1.3 Brain writing 6-3-5	143
Tool # 1.4 Bulletin Boards	143
Tool # 1.5 The Delphi Method	144
Tool # 1.6 Understanding the Topic	145
Tool # 1.7 Stimulating Creative Thinking	146
Tool # 2 Categorical Selections	148
Tool # 2.1 Pass-Fail Criteria	149
Tool # 2.2 Wants and Limitations Criteria	150
Tool # 2.3 Pareto Analysis	150
Tool # 2.4 Decision Matrix	151
Tool # 2.5 Pairwise Comparison Analysis	152
Tool # 3 Review Panel	153
Tool # 3.1 Expert Panel	153
Tool # 3.2 The Delphi Method	155
Tool # 3.3 Deliberative Forum	156
Tool # 4 Group Voting	158
Tool # 4.1 Multivoting	159
Tool # 4.2 Consensus	161
Tool # 5 The Problem Statement	162
Tool # 6 The Vision Statement	166
Tool # 7 The Scientific Model	169
Tool # 7.1 Is/Is-Not	172
Tool # 7.2 Four Windows	173
Tool # 7.3 Ishikawa Diagram	174
Tool # 7.4 Interrelationship Diagram	175
Tool # 7.5 The Gap Statement	177
Tool # 7.6 The Prediction Statement	178
Tool # 8 The Objective Statement	179
Tool # 9 The Research Concept	182

        Tool # 9.1 SWOT Analysis 184
        Tool # 9.2 Force Field Analysis 187
        Tool # 9.3 Win Conditions 188
        Tool # 9.4 The Research Project Skinny 189
    Tool # 10 Financial Analysis 192
        Tool # 10.1 Return on Investment 194
        Tool # 10.2 Net Present Value 195
    Tool # 11 Customer Assessment 198
        Example Customers 200
    Tool # 12 Selecting Research Projects 202
        Tool # 12.1 Sponsor Assessment 202
        Tool # 12.2 Success Zones 203

Citations and Notes 205
Further Reading 223
    Innovation 223
    Project Management Manuals 223
    Philosophy and Scientific Method 225
    Creative Thinking and Problem Solving 226
Acknowledgements 228
About the Author 229

INTRODUCTION

INNOVATION REQUIRES KNOWLEDGE that others do not have. Research projects explore the unknown in a quest to acquire knowledge about a scientific phenomenon. Research projects attempt to solve a problem, spark innovation and change our world. A researcher cannot explore our world without carefully planning where to go, without controlling how to get there and without clearly communicating knowledge to others. Research projects require a special kind of project management.

Research projects share many similarities with other projects and can use many of the same project management practices that have been successfully implemented in other professions. Projects have a work breakdown structure[1] that describes a sequential series of tasks to reach a defined goal. Research projects follow the scientific method. In its simplest form, the scientific method[2] is analogous to the work breakdown structure of a traditional project that builds something.
1) Conduct a literature review to summarize previous observations.
2) Propose a hypothesis that predicts the outcome of an experiment.
3) Conduct that experiment and collect data.
4) Compare the predicted with the actual data.
5) Support or reject the hypothesis.

A second similarity is that all projects are conducted by teams of skilled interdependent professionals who work together towards a common goal. Scientists join temporary research teams to conduct specific research projects and achieve specific goals that often include professional development. Once that research project is completed, the team disbands and another research team forms. This is common with graduate and postdoctoral students in universities, but also occurs in commercial and government laboratories as research programs address new problems.

In addition, all projects have a fixed budget and often a fixed duration. The stability, duration and amount of funding determine the scope and quality of a research project. All projects are temporary activities that have a defined endpoint.

Finally, all projects produce a deliverable for a customer. That deliverable is usually a tangible product but a research project produces knowledge. The customers of research projects may not be obvious but they use knowledge from research to inform, to decide, to solve problems and to innovate. Sometimes, the customer is the general public. Sometimes, the customer is a contractor paying a fee for a service. Sometimes, the customer is a teacher, student, businessman, entrepreneur, regulator or simply a curious person. All customers have needs and expectations that must be recognized to be successful.

Because of these similarities, many traditional project management tools work equally well for research projects, but research projects differ in several ways that mandate a different style of project management. In my perspective, management of a process to learn something is different than management of a process to build something. Research projects are not well suited to many of the traditional project management practices for several reasons.

First, research projects rarely follow the simple literal scientific method. There are iterations and then there are more iterations of the workplan. Some observations are unexpected. One experiment is never a conclusive test of a hypothesis. Many research projects do not even propose or test a hypothesis, but use an experimentalism philosophy that follows the data. Variation is everywhere obscuring logical conclusions. Long-term, logical planning of a sequence of experiments is futile.

Second, the research project includes a sequential series of decisions. Which of the many important and urgent scientific problems to address? Which hypothesis to propose? How to test the hypothesis? What observations and data to collect? How to compare predicted and actual observations? How to interpret the observations? Which experiment to do next? What result is good-enough to make a conclusion? The quality of these decisions is more important to the outcome and success of the research project than the scientist's technical skill.

Third, the observations, analysis and conclusions are made by biased people in an environment filled with paradigms that are skewed by their historical experience. Observations are subject to multiple interpretations. Different individuals have different perspectives, make different conclusions and derive different types of knowledge from the same experiment. The craft, art and skill of the scientist impact the methods used, the results obtained and the subjective conclusions made. The knowledge created by research projects is, therefore, fallible.

Fourth, the knowledge created by a research project is useless until it is effectively communicated and judged by others. The research will be subjected to intensive review, criticism and even ridicule by peers, by professionals, by competitors and by adversaries. Rigorous experimentation, clear communication and consultative decision-making are essential to pass this scrutiny.

Finally, research projects can be endless. There is always more to learn. There are always unanswered questions, unsolved problems, opportunities for innovation. Decisions must be made in the absence of information. Research is expensive and resource intensive, but has a fixed budget. Consequently, research projects must be focused, precisely defined, carefully planned and expertly controlled to achieve a targeted objective.

My experience is in the management of research projects in the life sciences, but I believe that the principles and philosophy of research project management can be applied broadly. I incorporated some research project management processes into my research projects, which helped me to manage those projects better. I found that being successful in research today requires more than knowledge of the "facts"; it requires management skills that must be learned, practiced, honed and polished.

My objective in writing this book series *Research Project Management* is to provide you with management tools to enable you to complete successful, innovative research projects that have global impact. I am confident that young scientists who are beginning their research careers may adopt some of these processes and tools more easily than others, but I learned many of them late in my career. Regardless of the stage of your career, you may be able to improve how you conduct research projects by incorporating some of these management processes.

I adapt several management concepts and merge these into the scientific method, including creative thinking, leadership, teamwork, project management and communication. The result is a management philosophy for research projects that I have summarized in ten principles. The following chapters provide the framework, introduce concepts and define the terminology that I use in research project management. In the next chapters I outline how these principles can be applied to the critical decisions that you as a skilled, well-trained researcher must make as you propose a research project for funding by a sponsor.

The selection of which problem to address is the most significant decision that a researcher can make. Some solve problems that allow them to make new products. Some solve problems affecting society, such as cure a disease or prevent starvation. Some solve a puzzle that sparked their curiosity. Some create a business opportunity in a new market. Others improve the efficiency of a running process. Regardless of your quest as a researcher, you will face four different categories of problems.

A. Problems that are poorly described.
B. Problems that are poorly understood.
C. Problems that have several hypothetical solutions.
D. Problems that need validation of a solution.

Different problems require different creative solutions. Close-ended problems have a single solution. Most scientific problems are open-ended and have many possible solutions; some are likely to be better than others. Adaptive problem-solving makes incremental changes and offers guaranteed results, including good scientific publications. Radical problem-solving has a much higher risk of failure but offers to create innovation, to change scientific paradigms and to change our world. Problem-solving requires a vision of what the world will be like without the problem.

Your vision of a solution will be based on scientific knowledge which is conceptualized in a scientific model. Without a validated scientific model, it is impossible to conceive a research project. The scientific model may be conceptual, or it may be a mathematical attempt to explain how things work. The scientific model is your roadmap to achieve your vision. If your model is sufficiently rigorous, you can proceed immediately to implement your vision. But, in most cases, there are gaps in your knowledge, assumptions that have not been validated, predictions that have not been tested, prototypes that have not been built. In these cases, you need to conduct a research project to achieve a very specific objective. The research project, however, needs funding. It needs a sponsor. Only the "best" proposals will receive financial support.

The foundation of any research project is, therefore, a problem that needs to be solved, a vision for a potential solution and a scientific model predicting how to implement the vision. These are encapsulated in a research proposal.

The last section of this book contains a set of tools, collectively called the Scientist's Toolbox[3], that provide recipes with a step-by-step instruction to guide you through each process as you develop your research proposal. These tools are written based on my experience using laboratory manuals to conduct laboratory analyses and are written in that mindset. Some of the tools have multiple variations and components that can be used in different situations. These tools are guides, not rigid prescriptions. Some will work for you in some situations, others will require adaptation. Feel free to experiment to find those that are best suited to your projects as you perfect your skills in research project management.

PART ONE

RESEARCH PROJECTS

EXPECTATIONS OF RESEARCH PROJECTS

As a research scientist, you expect to impact the future. You expect that your research will create knowledge, shape new technology, initiate innovation, change what we do and change how we do it. You expect that your research will improve society, people's lives, families and the environment.

Research means different things to different people. To some, research projects are conducted to satisfy their curiosity about the world around them. To others, research creates technological advances that improves our health, our environment and our lives[1]. In the terminology that I will be using, research projects address problems - a wide range of problems that may be either puzzles or adversities, that may be real or perceived, that may be social or economic, that may affect us today or in the future.

Research scientists are therefore problem-solvers. Research projects create knowledge in an attempt to solve a problem. In project management terminology, a problem is the gap between where you are and where you want to be. Some scientists want to solve a commonly recognized problem in society, cure a disease, prevent starvation or mitigate climate change. Others want to capture a business opportunity in a new market. Others want to answer a scientific question sparked by their curiosity. All seek to acquire more complete knowledge about a scientific phenomenon. All seek change. All contribute to a never-ending stream of innovations that change our society, our environment, our world.

Society, through its policymakers, expects you to understand its problems and to implement a significant change to solve those problems. People expect practical, real solutions to their problems, not abstract theories, not probabilities. They expect accurate results and they want them now. They expect immediate answers to today's problems, not next year, not in 10 years. They expect the facts to be clear, irrefutable and consistent. They expect security and safety; any risk, especially any risk related to their family's health, is unacceptable. These are high expectations[2].

Because of these expectations, your quest for knowledge and innovation will never end. Your research may reduce starvation; yet, people will crave a more diverse supply of meat, bread, fruit and organic vegetables. Your research may prolong people's lives and develop new treatments for disease; yet, epidemics and pandemics will still threaten to sweep across the globe. Your research may create innovative technology; yet, you must adapt to other's innovations and their disruptions to remain relevant.

A popular myth[3] is that accomplishments in science and their consequent innovations are achieved by an inspired lone genius. We envision a rugged individual working in solitude in a small laboratory, hidden from managers and administrators, supported by an anonymous benefactor, ignoring criticism and shunning interaction with peers. This myth might have applied to Charles Darwin in Victorian England, but today innovation is implemented by well-funded research teams[4] of talented, interdependent scientists. The teams merge complex scientific and technological information from diverse disciplines. In doing so, scientists rely on one another for encouragement, information, inspiration and technical expertise. You will need more than scientific facts and technical skills to be successful as a scientist. You must work well with others. You must communicate well with others. You must make joint decisions.

The culture[5] in many of our Research and Development (R&D) organizations fails to recognize this interdependence. You were trained in graduate school to be an independent scientist, with excellent technical skills. But were you also trained in the skills needed to work with other skilled scientists towards a shared goal? Were you trained in creative teamwork? Were you trained in group decision-making? Probably not.

You may want to follow your curiosity wherever it leads. You may believe, as many research scientists do, that planning research is futile because you cannot predict the outcome. Nonetheless, the directors who manage corporate finances and the politicians who spend our tax dollars are refusing to fund research projects solely to satisfy a scientist's curiosity. Consequently, government and organizational funding of basic research to acquire knowledge has been declining in most countries over the past decades. They are demanding more accountability in the research they sponsor and consequently, they are demanding that you use more project management.

You may be motivated to use better management because you experienced a crisis or because you foresee one in the future. Reviewers may have criticized your management skill. Your research grant may be smaller. You may plan to move to a new R&D organization. You may be seeking new sponsors.

Regardless of your motivation, my contention is that you will conduct a better research project if you use a set of research project management processes. These tools will help you with the logistics of research. However, research projects are different. The primary reason for using research project management tools is to facilitate communication. You must communicate your plans, progress and accomplishments to others in a manner that motivates, empowers and informs. Your exploration of the unknown not only requires planning where you want to go and controlling how you get there, but it requires the support of others. You must clearly and explicitly communicate with others to gain their support, to maintain their commitment and to make joint decisions.

CYCLE OF INNOVATION

KNOWLEDGE ABOUT A scientific phenomenon is created using the scientific method. In its simplest form, the method is a systematic approach that requires you to ask a question, to propose a hypothetical answer and then to conduct an experiment to test the validity of that hypothesis. In practice, the scientific method is much more complicated. The scientific method is fact-based, involving scientific models (or theories), hypotheses, experiments and empirical data, but it is also opinion-based, involving people's perspective, experience, bias and beliefs that guide them in making conclusions. Like others, I will argue that the scientific method is not a method at all but instead, it is a philosophy.

From another point of view, Sawyer[1] teaches that problem-solving occurs in a series of steps:

1. Ask the right question.
2. Acquire the technical knowledge on the subject.
3. Incubate the problem and allow your subconscious to contribute.
4. Generate lots of ideas.
5. Merge and combine ideas.
6. Select the best ideas.
7. Get going and do something to implement your great idea.

Sawyer's Zig and Zag process is quite compatible with the scientific method. I will be advocating variations of a merged creative process throughout the following chapters.

> *Sawyer: No matter what kind of creativity I studied, the process was the same. Creativity did not descend like a bolt of lightning that lit up the world in a single brilliant flash. It came in tiny steps, bits of insight and incremental changes. Zigs and zags. When people followed those zigs and zags, ideas and revelations started flowing.*

To begin, recognize that there are four different types of problems that scientists seek to address. These problems ask different questions, require different creative solutions and therefore, require different categories of knowledge to solve.

    A. Problems that are poorly described require more observation and information.

B. Problems that lack understanding require better scientific models.
C. Problems that have several hypothetical solutions require testing.
D. Problems that need validation of a solution require a prototype.

Research projects have traditionally been classified simplistically as either basic or applied, classified as such based on the problem being studied and the researcher's stated objectives to sponsors. From my experience, I found that the management of basic and applied research projects is fundamentally the same. Similar management process are used because both seek knowledge. The problems may be defined differently. The objective may stated differently. The acquired knowledge may be applied differently. But the management processes to acquire that knowledge are the same.

Instead of using the terms basic and applied to describe research projects, I am proposing a classification for research projects based on the design thinking model[2]. In my Cycle of Innovation model, research projects address the four different categories of problems listed above. The roles that you may play in your quest to solve those problems are different. I will call these scientists: Observer, Dreamer, Explorer and Builder, respectively.

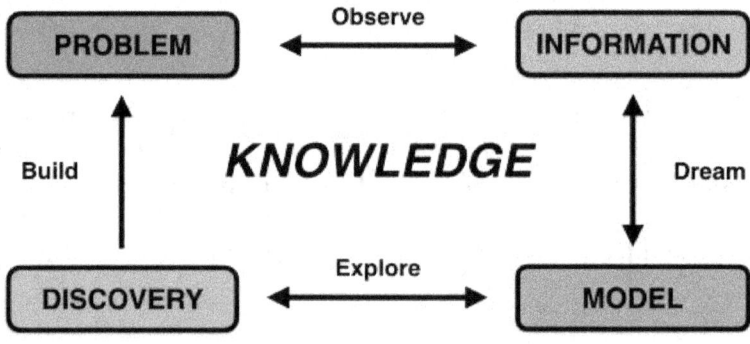

The Observer develops an understanding of the problem by collecting data and making measurements of the associated scientific phenomena; the Observer creates information.

The Dreamer molds information into relationships that make

predictions; the Dreamer creates a scientific model.

The Explorer tests the hypotheses proposed by a scientific model using different treatments and formulations; the Explorer creates discoveries and inventions.

The Builder uses knowledge in all of its forms - information, scientific models, inventions and discoveries - to construct prototypes that enable practical solutions to the original problem; the Builder creates an innovation.

My Cycle of Innovation model is overly simplistic because the acquisition of knowledge is an iterative process, not a gated linear process with clear distinctions and hand-offs. New scientific models suggest new measurements that can be made. Discoveries suggest revisions to improve scientific models. There is no Standard Operating Procedure or Six Sigma process[3] to acquire knowledge. The roles of a scientist are fuzzy with individuals playing Observer, Dreamer, Explorer and Builder roles at different times and in different situations during their career, or even during a research project. Nonetheless, my model is useful to identify the skills and management tools that you will need to solve different categories of problems. Your roles require different management skills just as much as they require different scientific skills. You must sometimes lead, sometimes work in teams and sometimes hand-off.

KEY POINTS

SOCIETY EXPECTS YOU as a research scientist to understand its problems and to implement a significant change to solve those problems.

Accomplishments in science and their consequent innovations are achieved by well-funded research teams of talented, interdependent scientists.

Government and organizational sponsors of research are demanding more accountability from their funding and consequently, they are demanding that you use more project management.

The primary reason for using research project management tools is to facilitate communication among scientists, managers and decision-makers.

There are four different types of problems that scientists seek to address. They ask different questions, require different creative solutions and require different categories of knowledge to solve:
   A. Problems that are poorly described require more observation and information.
   B. Problems that lack understanding require better scientific models.
   C. Problems that have several hypothetical solutions require testing.
   D. Problems that need validation of a solution require a prototype.

In my Cycle of Innovation model, scientists play four different roles: Observer, Dreamer, Explorer and Builder, who address the four different categories of problems. The Observer develops an understanding of the problem; the Observer creates information. The Dreamer molds information into relationships; the Dreamer creates a scientific model. The Explorer tests the hypotheses proposed by a scientific model; the Explorer creates discoveries and inventions. The Builder uses knowledge in all of its forms to construct prototypes; the Builder creates an innovation.

These roles require different management skills as well as different scientific skills. You must sometimes lead, sometimes work in teams and sometimes hand-off.

PART TWO

RESEARCH PROJECT MANAGEMENT

THE WORKPLAN

IT IS OBVIOUS from the preceding discussion that research projects are conducted for a variety of reasons and encompass a spectrum of project designs. The distinction among Observation, Modeling, Discovery and Development projects is fuzzy. Hybrid projects that address a problem using more than one approach are common in many R&D organizations. Although the administrative organization and terminology differ among R&D organizations, Observation, Modeling, Discovery and Development projects have several common components that must be addressed by research project management. Each must:

1. Define a problem. Ask the right question.
2. Envision an opportunity that gives a potential benefit.
3. Understand current knowledge in the form of a scientific model, identify gaps in knowledge and make predictions.
4. Establish an objective to learn something.
5. Enlist and maintain support of others.
    a. Administrators and managers who control key resources.
    b. Peers who will provide advice.
    c. Service providers.
    d. Customers who will use the deliverable.
    e. Sponsors who will provide the funding.
6. Recruit and organize a research team that is motivated to explore the unknown.
7. Develop strategic and tactical plans on how to achieve the objective, design experiments and conduct those experiments.
8. Make scientific, management and logistical decisions.
9. React to unexpected events that pose risk or offer opportunities.
10. Communicate knowledge as a deliverable to a customer.

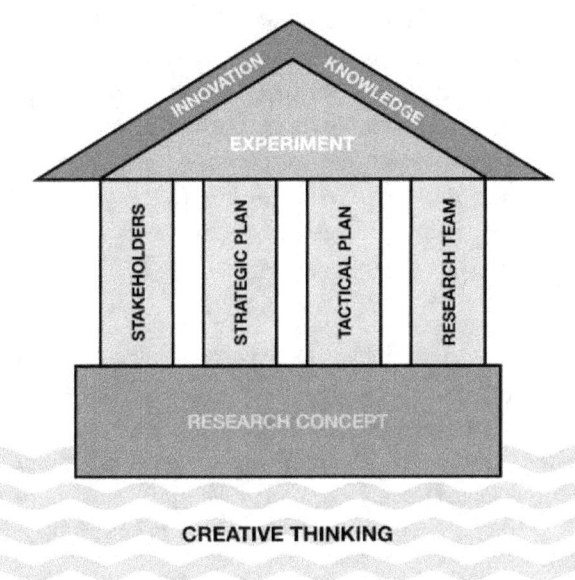

The cartoon depicts the components of research project management as being analogous to a house. The analogy is equally applicable to Observation, Modeling, Discovery and Development projects, as well as hybrid projects. The deliverables (knowledge and innovation) are produced from experiments (data and information gathering activities) that rely upon four pillars of support - stakeholders, strategic planning, tactical planning, research team. These pillars rest on the firm foundation of a research concept that defines a problem, envisions a change, provides a scientific model for a solution and states an objective. The research concept is rooted in creative thinking. If any of these pillars fail, the research project will collapse. If the foundation of a research concept is weak, the pillars cannot support the execution of the project. If the thinking fails to be creative, imaginative and rigorous, the foundation will shift, and the project will collapse. Effective communication is the glue that holds all of these components together, fills the cracks and prevents collapse.

All of this research project management occurs within some form of R&D organization. Some are primarily academic including universities, public research centers, non-profit and government laboratories. Others are divisions of a commercial for-profit business. Others are private sector, commercial laboratories that provide services under contract, again for-profit. Still others are start-up companies that develop new technologies for sale to larger organizations. The specific mandate of these R&D organizations is diverse. Some of these R&D organizations publish their knowledge in scientific journals and in patent applications. Other R&D organizations use their knowledge to establish regulations and guidelines to protect our environment and our health. Other R&D organizations acquire knowledge to teach students. Other R&D organizations apply knowledge to create commercial technologies and products but keep that knowledge secret.

> *Mordechai Ben-Ari[1]: Modern scientific work demands specialized knowledge and resources that can only be obtained in universities and research institutes set up for that purpose.......The day is long gone when a talented amateur like Michael Faraday could make important discoveries using homemade apparatuses. To make a new discovery in experimental science requires sophisticated and expensive equipment.....*

In spite of their different mandates, R&D organizations are quite similar. They develop, conduct and coordinate research projects to create knowledge. They manage the distribution and utilization of that knowledge. They have physical facilities and infrastructure that must be managed, maintained and allocated to projects. They have relationships with external groups and other R&D organizations that require careful communication. They employ creative technical experts who preserve the organization's knowledge. They have responsibility to manage, train, motivate, empower and recognize their scientists. Although some are better than others, they all use some form of research project management.

PROJECT MANAGEMENT

PROJECT MANAGEMENT IS a profession with over 800,000 accredited members in 210 countries worldwide[1]. The Project Management Body of Knowledge[2] details the accepted practices for a complete range of project management activities. Originally, the focus was on construction-like projects, their schedules and costs. Practices have been added subsequently for creative projects, such as software development.

Many R&D organizations are adopting the traditional project management practices in an effort to improve the quality, efficiency and cost-effectiveness of research projects. For example, the Center for Disease Control has developed strict project management practices for all of its projects. The CDC Unified Process[3] "... is a collection of practices, processes, tools, artifacts, and information that any project can use to structure, track, and manage activities and deliverables to better deliver high-quality investment outcomes on schedule, budget, and within scope." The UP informs and supports project managers and project teams in using the best practices, emphasizing quality.

James Lewis[4]: Project management is the planning, scheduling and controlling of scarce resources to achieve desired results.

Other R&D organizations are also trying to improve their research projects by incorporating project management practices into their portfolio and into their culture. However, I found that when I tried to apply these practices to managing my research projects, I had problems. My colleagues, my research team and my managers were resistant, sometimes for good reasons. My research team viewed project management as a rigid, time-consuming, bureaucratic process that hindered their daily work without tangible benefit. My managers were reluctant to delegate decision-making authority due to concerns about control, responsibility and accountability. Nonetheless, I believe that many project management practices are simply good management tools that need to be adapted to fit into the scientific method and the creative culture of a research team[5].

The first challenge that I encountered when trying to implement project management into my research projects was the lack of a common vocabulary. All scientists employ the scientific method in their research projects, and they are familiar with that terminology. In contrast, the definitions in project management vary among different R&D organizations, especially if other divisions in the organization do construction projects. It becomes confusing when project management terms are translated into research projects that traditionally have used the same terms to mean something different. As I will emphasize, communication is at the core of research project management and to communicate effectively we must have a common understanding of the words and terms that we are using. I start with some basic definitions that will allow me to frame my subsequent recommendations.

PROJECTS AND PROGRAMS

The standard definitions of a project and a program from the Project Management Institute[6] are:

Project Management Institute definition of a project: A temporary endeavor undertaken to create a unique product, service, or result.

Project Management Institute definition of a program: A group of related projects, subprograms and program activities that are managed in a coordinated way to obtain benefits not available from managing them individually.

As Brown and Hyer[7] point out, the definition of a project is exceedingly broad. A project may be conducted to change processes, create products or improve services. It may be small (organize a party) or large (build a space station). Similarly, my definition of a research project encompasses all levels of complexity, size or importance. A graduate student's thesis research is viewed by the graduate student as their research project. The major professor, who is managing a National Science Foundation grant supporting several graduate students and postdoctoral fellows across 3 institutions, considers this to be a research project. Others may consider an effort as large as the Human Genome Project to be a research project.

I make a clear distinction between research projects and research programs because their management, their planning and their deliverables are different. My definitions slightly modify those from the Project Management Institute:

Research Project. A temporary endeavor with a defined beginning and end undertaken to create knowledge.

Research Program. A set of inter-related coordinated research projects undertaken to solve a problem.

Research projects are temporary, have a defined beginning, have a defined end point and deliver knowledge to a customer. The deliverable from an Observation, Modeling or Discovery project is knowledge in the form of intellectual property, a scientific publication or a technical report. The deliverable from a Development project is knowledge in the form of tangible prototype for a commercial product, technology or process.

In contrast, a research program has a strategic objective related to a problem associated with a scientific phenomenon. Several Observation, Modeling, Discovery and Development projects may occur in the same research program, either concurrently or sequentially. After one research project is completed, another project is likely to continue as part of the same research program. More can always be learned; more experiments can always be conducted. Problems are never completely understood; they are never completely solved; improvement is always possible. New and improved products are always needed. For example, a R&D organization may have a research program on rheumatoid arthritis, composed of several different approaches to gain understanding and progressively better treatments. Another R&D organization may have a research program developing maize hybrids that releases a new hybrid every three years. Research programs are ongoing and end only when the strategic objectives of the sponsor change.

I belabor the distinction among projects and programs because it is a fundamental distinction for research project management. Some R&D organizations merge the projects of the Observer, Dreamer, Explorer and even the Builder into one large hybrid project. They use project management tools, when they should be using program management tools. They confuse tactics and strategy. These "projects" are typically governed by a senior executive who provides direction, oversight and decision-making, but who is unable to plan, monitor or control the experiments. Consequently, agile and flexible management are impossible to implement because decision-making by senior executives is centralized and slow; tactical changes require senior executive approval; disruptive budget fluctuations are common.

Most senior managers more appropriately focus at the level of the research portfolio, which consists of many research programs that support the R&D organization's mission. The senior executives in commercial R&D organizations review and adjust their portfolio at least annually to meet their changing priorities. Universities are slower to react but adjust their portfolios through the appointment of new faculty or the creation of institutes and other administrative structures.

Project management practices can be used by individuals, teams, agencies, startup companies, businesses and multinational corporations. Many professional groups have revised the traditional project management practices to meet their needs better. Some of these revisions are especially relevant to research projects. The following are a few highlights.

AGILE PROJECT MANAGEMENT

The Chaos Report in 1994[8] from the Standish Group studied the high rate of failure and cost overruns of software development projects. In response, the project management community developed alternative ways of managing creative projects. These improvements led to the concept of agile project management[9]. Agile project management is an umbrella term that has been applied to include several different approaches. The common practice is that the deliverables are submitted in stages to customers as prototypes for feedback, revision and improvement. The projects becomes a time-boxed incremental process that is highly interactive between a project team and its customers. The agile project follows a repetitive planned cycle:

Design → Build → Test → Release → Design → Build → Test → Release →

Agile project management for software development is guided by a set of written principles, Twelve Principles of Agile Software and the Agile Manifesto[10]. The Agile Leadership Network developed a Declaration of Interdependence to serve as the guiding principles to employ agile and adaptive approaches for linking people, projects and value[11]. The statements provide a view of managing complex and uncertain projects that is particularly relevant to the management of research projects and research teams. I have borrowed liberally from these principles in writing the principles and the tools used in research project management.

Most researchers already use a form of agile management when reporting a series of experiments in scientific publications. They obtain feedback from reviewers and peers (arguably their customers) before moving on to the next set of experiments. Consequently, agile project management practices and principles can be readily adapted into research project management.

ADAPTIVE PROJECT MANAGEMENT

Adaptive management[12] is a structured and systematic process for continually improving decisions, management policies and practices by learning from the outcomes of decisions previously taken, or in other words "learning by doing". This concept has been applied to project management when the deliverable is poorly defined, when the customer is unidentified or in other situations that create uncertain requirements for the deliverable. The essence of adaptive management is captured in this motto:

4-H Canada Motto[13]: Learn to do by doing.

Two principles are common to agile and adaptive project management. Both require iterative decision-making based on learning from experience. Both require flexibility and the ability to change tactics or even strategies. The original concept of adaptive management also required quantitative analysis to make better decisions based on actual data, such as statistical analysis, hypothesis testin, performance measurement and quantitative project risk analysis.

Adaptive project management is a common practice in research projects. Most research programs cycle among the strategizing, planning and executing stages. After one experiment is completed, another is planned based on its results. The research team learns as it conducts experiments. Each experiment creates more technical knowledge, tests alternative hypotheses and revises scientific models. Nonetheless, some experiments fail; some projects fail; some strategies fail; unforeseen opportunities arise. Plans change.

Brian Greene[14]: Exploring the unknown requires tolerating uncertainty.

Lev Virine[15] summarizes the philosophy of adaptive project management and gives advice on how it may be used in a project, which seems especially relevant to research project management:

- Plan a project that has strategic flexibility.

- Plan a project that has multiple phases or iterations.

- Make a detailed plan only for the next phase or iteration of the project. Future phases or iterations may change based on the outcomes of previous iterations.

- Determine what will happen to the project if certain risks occur. Consider risk when choosing a strategy.

- Continuously measure actual results against the original plan.

EXTREME PROJECT MANAGEMENT

Doug DeCarlo[16] proposed a radical shift in the philosophy of project management. DeCarlo contended that project management practices were developed for ideal projects. In an ideal project, the project manager receives clear instructions from the customer as to what is wanted, when it is wanted and how much the customer is willing to pay for it. These requirements never change. The methods for achieving the results are proven and well understood. The technology is well established. The results are implemented easily and provide incremental benefit to everyone. No one opposes the project.

This ideal project does not exist. Many projects have customers with conflicting interests, have no sponsor, use new unproven technology and meet passive (even active) resistance. People are assigned to multiple projects at the same time, and consequently lack commitment. The deliverables and resources change overnight. The deadline is unrealistic. The work environment is chaotic.

DeCarlo called these eXtreme projects. He concluded that, in these projects, a linear approach to project management fails. The standard tools, templates and processes of traditional project management are almost useless.

DeCarlo's eXtreme project management is an integrated set of principles, values, skills, tools and practices that work under conditions of rapid change and uncertainty.

- Whereas traditional project management imposes standard processes on people, eXtreme project management makes the process serve people.
- Whereas traditional project management centralizes control of people, processes and tools, eXtreme project management distributes control.
- Whereas traditional project management is about taking charge of the world, eXtreme project management is about taking charge of yourself and your approach to the world.
- Whereas traditional project management is about managing, eXtreme project management is about leading.

Research projects have many of the characteristics of DeCarlo's eXtreme projects and I have adapted many of his recommendations for managing projects into research project management.

RESEARCH PROJECTS ARE DIFFERENT

I WILL BE advocating throughout this book series that research project management tools have tremendous potential to improve the efficiency and the quality of research projects. However, the cookie-cutter implementation of traditional project management practices in R&D organizations will fail to provide the desired benefits to creative processes. Research projects are different from construction projects. Traditional project management practices were designed to control schedules and costs. Research project management practices are designed to facilitate communication and decision-making. This section introduces some of the differences that require special attention.

> *Moore and Shangraw[1]: . . . most project management strategies were designed for business, not science . . . .*

Knowledge is the Deliverable. Research projects explore the unknown. Research projects do things that have never been done before. All research projects deliver scientific knowledge to customers, but knowledge is an intangible, poorly defined deliverable. Scientific knowledge is expressed in the form of scientific models that attempt to explain a phenomenon in progressively better and better ways[2].

Knowledge is based on empirical observations and data. The interpretation of data may be biased, and consequently knowledge is skewed in the favor of that bias. Data may be open to multiple interpretations. Knowledge may create controversy and ethical dilemmas. Multiple customers may have conflicting needs and expectations. The end point of a research project is not clear. More can always be learned. These features conflict with the traditional project management philosophy of a deliverable.

Knowledge must be communicated to its customers to have a benefit. Knowledge from research projects may be published in scientific journals or protected as intellectual property or sometimes both. Patent applications are a common means of protecting intellectual property and require careful documentation of what was done, by whom and when. Trade secrets are closely guarded. Research projects must therefore include a communication plan on how knowledge will be transferred to its customers and how any potential intellectual property will be protected.

The concept of a customer for knowledge is fundamental to research project management. The customers of research projects use knowledge to inform, to decide, to solve problems and to innovate. Sometimes, the customer is the general public. Sometimes, the customer is a contractor paying for specific knowledge. Sometimes, the customer is a teacher, student, businessman, entrepreneur, regulator or simply a curious person. Sometimes, the customer may be the legal department or marketing unit of a commercial organization. Sometimes, the customer is a team member or a colleague who will use the product of your experiment/task as an input in their experiment/task. In the latter case, knowledge (and often tangible material) is passed from one research team member to another as part of the project's workplan. The results of one experiment inform others how to design the next experiment, how to impose treatments and how to make measurements. The research team is its own customer as it learns and explores the unknown.

Quality is Subjective. Quality is of course important to any project, but quality in a research project is more difficult to define or to measure than the quality required in a construction project, or even a software development project. Quality of knowledge is a perception. A research project will always produce knowledge, provided it is technically correct. A research project may fail to produce the commercial product that the customer expected, but it will always produce knowledge. Bob Gundlach[3] commented that failure is an important step in innovation and noted that many inventions come after initial failure.

Roman Szpur[4]: Learning why an idea fails is just as important as knowing why it works.

Some assume that publication in a peer-reviewed scientific journal is certification of high quality. The knowledge created by a research project is ideally robust, but this ideal is not always achieved in published research reports. For example, the 2005 publication of John Ioannidis in PLoS Medicine labelled many previous medical studies as irreproducible.

John Ioannidis[5]: ... it is more likely for a research claim to be false than true.

Some assume that creation of intellectual property is certification of high quality. The invention described in a patent meets certain requirements, including novelty and non-obviousness, that impose a relatively high standard of quality. The description of the invention must also meet standards of disclosure sufficient to allow someone skilled in the art to repeat the invention. However, a patent application, even when published, has yet to be scrutinized by patent examiners. In that manner, it is similar to a manuscript submitted for peer-review.

Quality of a research project can be measured by the impact that the knowledge has on a scientific model and subsequent research. The number of citations, the impact factor of journals, h-index and i-10 index are all quantitative scores of how your peers value your research publications[6]. Note that Gregor Mendel's research was ignored for almost 40 years. Payton Rous was awarded the 1966 Nobel Prize for his work on retroviruses in 1910. Needless to say, the quality of knowledge, and therefore of research, is a subjective criterion.

Eyre-Walker and Stoletzki[7]: ....the number of citations a paper accumulates is a poor measure of merit and we argue that although it is likely to be poor, the impact factor, of the journal in which a paper is published, may be the best measure of scientific merit currently available.

Another unique aspect of a research project is that the interpretation of the data in scientific publications is subjective, depending on the bias, perspective and motives of the authors. As a result, the interpretation and the perceived quality of the research project vary with the context.

Noreena Hertz[8]: All of us show bias when it comes to what information we take in. We typically focus on anything that agrees with the outcome we want.

From a project management perspective, the quality of your research project will be determined by the satisfaction of your customers and stakeholders with the knowledge that you have created. Their assessment of quality is based on perception. Their perception is based on how well you have communicated with them to convey your accomplishments, in the context of their expectations.

Conflict is Inherent. Creating knowledge is competitive, controversial, judgmental and subjective. Knowledge can create innovations that have good and bad effects depending on one's perspective. Society chooses whether the benefits outweigh the harmful consequences. Society's choices may be illogical based on emotional reactions. Research projects will be subject to criticism and that criticism may have no scientific basis.

Criticism and disagreements are also expected from competitors and even from some stakeholders. Conflict can also occur within research teams. When properly managed, professional conflict within a research team stimulates creativity. Conflict within a research team creates new paradigms, overcomes complacency, corrects mistakes and recognizes opportunities. Conflict prevents Groupthink[9] and leads to better decisions. Managing creative conflict in a team environment to create innovation, instead of disagreement and dysfunction, requires special skills, mutual professional respect and a creative culture[10].

Robert A Heinlein: I never learned from a man who agreed with me.

Personal conflict is often the result of misunderstandings. Misunderstandings are the consequence of poor research project management. Clear communication of unambiguous goals, plans and expectations among the members of a research team is the most effective way to minimize personal conflicts.

Progress requires Decisions. A major distinction between a research project and a construction project is that a research project is composed of a prescribed series of decisions. The research team must decide which scientific model to accept, which hypothesis to test and how to test it experimentally. Once the experiment is completed and the data collected, the research team must decide whether the data are valid or if technical problems require the experiment to be repeated to correct those problems. If the data are considered valid, the research team must decide whether to accept or reject the hypothesis. Then the research team must decide whether to revise the scientific model based on the hypothesis decision. Alternatively, the team may need to select among a range of alternatives, formulations or designs. Finally, the team must decide how to communicate the results to the customers and to the public, giving special consideration to intellectual property.

Doug DeCarlo[11]*: The only reason to establish a team is to make decisions and solve problems.*

Unlike a construction project, the strategy of a research project embeds decision points into the workplan forming a decision tree with multiple alternative branches after each decision. A research project's workplan is not a blueprint. Long-term tactical planning is futile because the number of alternative paths increases exponentially at each branch-decision-point. The decision of which path to follow impacts resources and budgets. Consequently, decision-making, flexibility and adaptability are inherent features of all research projects.

The quality of the decisions made by the team determines the research project's success and the quality of the knowledge created, more than the technical skill of the scientists doing the experiments, more than the quality of the research facilities and more than the size of the research grant.

Constraints Force Assumptions and Compromise. In project management terms, scope describes the boundaries of a project. Scope defines what the project will do and what it will NOT do to provide deliverables to a customer. The scope of all projects is limited by three common constraints:

1. Time
2. Cost
3. Quality

The scope of a project is constrained by the time allowed, its budget and the quality required. Theoretically, only three of the four variables can be fixed; the fourth must float and is defined by the other three; all four variables cannot be fixed. This is known as the "triple constraint" in project management terminology[12].

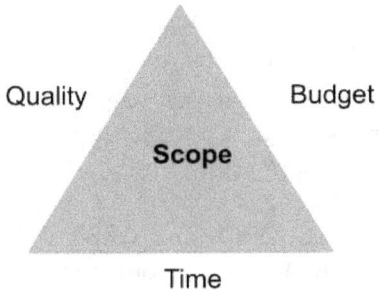

In a research project, two or three of these constraints are commonly fixed in different types of R&D organizations. Research projects at universities are funded by grants with fixed budgets and fixed durations; this means that only scope and quality can vary. Research projects at government agencies that set regulations or oversee public programs must meet exacting quality standards that can withstand public scrutiny. These agencies conduct research projects in which scope and quality are fixed, whereas budget and duration may be flexible.

Research projects in commercial R&D organizations or in academic R&D organizations under contracts are under competitive pressure. The first to patent or the first to the market wins. These R&D organizations conduct research projects that must be completed quickly with sufficient quality to merit patenting or marketing. Consequently, time, scope and quality are fixed, whereas budgets may be more flexible.

Other constraints limit which experiments can be done and, therefore, limit the scope and quality of a research project. These constraints must also be considered when planning and conducting a research project. Legal, regulatory, ethical, safety and infrastructure constraints are specific to each scientific discipline and R&D organization. There are regulations and guidelines on how construction projects are conducted. Likewise, there are regulations and guidelines on how research projects should be conducted.[13] United States Environmental Protection Agency regulations prescribe various procedures and designs for research that may impact the environment in an effort to protect the subjects, the researchers and the environment.[14] The Animal and Plant Health Inspection Service (APHIS) in the United States Department of Agriculture (USDA) regulates the testing of genetically modified crops under its permitting, notification and deregulation procedures. In most fields of medical science, experimentation is performed first on laboratory animals, but these experiments also have constraints and regulation[15]. Ethics and regulation[16] often demand that the scientific method be compromised. Sometimes, assumptions in a scientific model must be accepted without rigorous experimentation or validation.

Probably the most significant constraint to a research project, although we often fail to admit it, is our own individual ability. You may lack the scientific skill, the technical knowledge or the perspective to create a unifying scientific model, to identify the critical hypothesis or to design the best strategy to test a hypothesis. In which case, you will require talented colleagues to complement your abilities and only jointly can you reach your full potential. Your research project may be constrained in its ability to explore the unknown if you cannot recruit excellent colleagues as research team members and collaborators. Three factors, which DeCarlo[17] calls "win conditions", will influence a scientist's choice of which research project to conduct and which research team to join. These also are potential constraints that influence the selection and design of research projects:

1. Stakeholder and customer satisfaction. Often research projects are constrained by the expectations of the customers who will benefit and use the knowledge created and by the stakeholders who manage the resources, including funding, required to complete the project. Their satisfaction is based first, on understanding their expectations and second, on communicating your plans and your accomplishments effectively to them.

2. Team satisfaction. Your research team must gain experience and skill through their participation. There must also be tangible benefits in the form of scientific publications that will promote the careers of individuals. The greater the impact of the publication is, the greater the reputation of the research team or "laboratory" is, and the greater the prestige of the R&D organization is, the more likely your career will benefit. This is true for graduate students, postdoctoral fellows

and established researchers.

3. Return on Investment. Sponsors expect benefit, often in the form of economic return for their financial support of research. Many research projects are not conducted because their potential ROI fails to meet the sponsors expectations. Still others are deemed to fail due to cost overruns, delays and changes in market conditions.

Uncertainty and Variability are Everywhere. Risk is a factor in all projects[18]. According to Murphy's law[19], some things just go wrong. But a higher degree of uncertainty exists in a research project than in any other type of project because some of the basic assumptions in the scientific models may be false. In many scientific disciplines, notably biology, the quantitative data obtained from an experiment are variable. Variability in data may obscure relationships. Conclusions in biology are based on probabilities, not certainties. Experiments use statistics to uncover relationships that require subsequent verification and confirmation. Both false positive and false negative conclusions are easily made. Budget and time constraints may exclude confirmation experiments. Empirical data are open to interpretation and potentially to bias in that interpretation. This degree of uncertainty and variability is intolerable in a construction project but is inherent in all research projects.

Yet, an unexpected result may pose an opportunity. If the observations from an experiment are inconsistent with a hypothesis, the scientific model may be incorrect. This is not project failure but an opportunity to refine the scientific model, to create a new paradigm and to create intellectual property. The invention described in a patent application must not only be useful, but it must also be novel and not be obvious to someone "skilled in the art". If the existing scientific model predicts that something will happen, then the discovery cannot be considered novel nor not obvious.

> *Enrico Fermi[20]: ... if the result confirms the hypothesis, then you've made a measurement. If the result is contrary to the hypothesis, then you've made a discovery.*

The research team must be vigilant enough to recognize these opportunities, disciplined enough to ignore rabbits posing as opportunities and empowered enough to change plans when a real opportunity arises.

Smaller is better. Research project management requires both strategic and tactical management. The advantage of organizing research into small, independently-managed projects is that their tactical management is greatly simplified. Several Observation, Modeling, Discovery and Development projects with distinct tactical objectives may run either concurrently or sequentially in a research program to achieve strategic objectives. Each has a distinct deliverable – collect information, build a scientific model, test hypotheses and build a prototype, respectively. Each may have a different Principal Investigator. Each may require a different team with different skills. Each may require different resources and infrastructure.

This is a fundamental concept in adaptive project management. For projects involving creativity and uncertainty, Lev Virine[21] recommended that both a project strategy and high-level project plan be prepared to enable flexibility. He recommended to split a project into multiple phases or iterations and define a more detailed plan for only the next phase or iteration of the project. Long term plans for future iterations are likely to change.

Organizing a research program, which has strategic objectives, into small, independent research projects, which have tactical objectives, has several other advantages compared to organizing the same work as one large hybrid project:

- The objectives and deliverables of each project can be defined more precisely.
- There are fewer decision and branch-decision-points in the workplan. This enables traditional project management tools to be used to create a work breakdown structure, identify a critical path, schedule work, assign responsibilities and create budgets.
- Each research team can be empowered to make its own decisions about the tactical approach, the work plan and day-to-day operations. This builds ownership and commitment in scientists.
- High risk options can be quickly explored in small projects using customized tactics and focused approaches.

- "Fast failures" can be terminated without disruption to other projects.
- Lessons can be learned as projects are completed and applied to improve the scientific or managerial quality of subsequent projects.
- "Small wins" can be achieved and incremental benefits can be delivered to the customer.
- The perception of success is created as each research project is closed successfully.

THE TEN PRINCIPLES OF RESEARCH PROJECT MANAGEMENT

MY THESIS IS that research projects require a different type of management than traditional construction projects because they have different processes, organization, deliverables and customers. Or in other words, management of a process to learn something is different from management of a process to build something. Research project management provides not only detailed planning, scheduling and budgeting, but more importantly, explicit communication and transparent decision-making processes. Effective communication gives transparency and understanding to the decision-making process. The research team that understands the vision is empowered to make decisions and motivated to persist in their exploration of the unknown. The decisions and choices that a research team makes will undoubtedly be scrutinized and criticized, that is simply part of science. That scrutiny will establish the true value of the knowledge created by the research project.

> *Research project management: Management of talented, creative people to acquire knowledge.*

Research project management is a management philosophy, not a management process, that facilitates communication and understanding. The philosophy provides motivation for scientists to explore the unknown, embraces communication among all participants, and enables group decision-making with a goal to create understanding of a scientific phenomenon. Research project management merges several best practices from traditional, agile, adaptive and eXtreme project management into the scientific method mixed with creative thinking and teamwork. The philosophy is summarized in these ten principles:

1. **Stimulate creativity.** Research project management requires problem-solving skills and creative thinking.
   Use divergent thinking to conceive many ideas and alternative approaches.
   Use convergent thinking to select the best ideas and options.
   Use strategic thinking to envision a goal that inspires others.
   Use tactical thinking to plan, schedule and control experiments.
   Use abductive logic to envision scientific models.
   Use deductive logic to propose hypotheses based on those models.

Nurture ideas.
Embrace and support the ideas of others.
Encourage and mentor others.

2. **Satisfy customer's expectations.** Research project management requires a recognition of customers and their expectations.

Identify others on your research team who rely upon you to produce high quality results on time to enable their work.

Identify all of the customers who will use the knowledge created by your research project to inform, to decide, to solve problems and to innovate.

Understand your customer's problems, envision opportunities and commit to developing practical solutions to their problems.

Achieve success by meeting your customer's expectations.

Define the specifications for each deliverable in quantitative, transparent terms to avoid misunderstandings. Then strive to exceed those expectations.

Interact closely with your customers throughout the project to ensure their support for workplans, decision criteria and selections.

Communicate with your customers in a timely manner. Misunderstandings, rumors and misinformation erode confidence and trust.

Strive for frequent, continuous delivery of reliable and rigorous knowledge.

3. **Build a strong research team.** Research project management recognizes that the best research teams are highly skilled, interactive and self-organizing. A research team must be motivated to achieve excellence.

Recognize interdependence within the research team.

Facilitate collaboration, cooperation and coordination.

Empower individuals and teams to make tactical decisions.

Ensure that everyone understands the objective, strategic goals and success criteria of the project.

Ensure that everyone understands the vision, scientific models, assumptions and hypotheses being tested.

Recognize and use the diversity of technical skill, experience, beliefs and thinking styles in your research team.

Provide a sense of meaning and purpose to each individual on your team.

Share responsibility and authority.
Support others and trust them to get their job done.
Enable others to make a difference.
Require accountability for results and shared responsibility for team effectiveness.
Stimulate and embrace professional conflict but avoid dysfunctional personal conflict.
Reward individuals and teams for creativity and accomplishments, not solely collegiality.

4. **Communicate effectively.** Research project management requires clear communication within the team and with all of the project's customers and stakeholders.

   Ensure that the knowledge created in your research project is understood, reviewed, accepted and used.

   Identify the stakeholders in your research project and define their expectations.

   Use a communication plan. Carefully plan, monitor and control WHO communicates WHAT to WHOM; avoid rumors and misinformation.

   Communicate openly and honestly, about your project's plans, progress, work products, and issues.

   Encourage everyone to speak about the good, the bad and the ugly without fear of reprisal.

5. **Be Explicit.** Research project management requires explicit statements that can be easily communicated and understood by all stakeholders.

   Explicitly state your research project's objectives and its criteria for success.

   Explicitly state your scientific model, including its assumptions, knowledge gaps and predictions.

   Provide clarity in your plans, progress and accomplishments.
   Clarity enables focus and coordination within the research team.
   Clarity motivates teams to persist to overcome obstacles.
   Clarity empowers teams to make decisions and selections.

6. **Keep it simple.** Research project management requires scientific models with the fewest assumptions and management processes

with the fewest policies, fewest standard operating procedures, least bureaucracy and greatest flexibility.

Use the Principle of Parsimony (Occam's Razor) frequently.

Make your scientific models as simple as possible.

Make your research project management processes as simple as possible.

Keep each research project relatively small, with specific tactical objectives.

Use several research projects as modular components of a research program to achieve your strategic objectives.

Adopt agile and adaptive project management.

7. **Plan carefully.** Research project management requires both strategic and tactical planning.

Focus on producing rigorous knowledge, not rigid plans.

Define explicit success criteria and then plan accordingly.

Use traditional project management practices for the tactical planning, monitoring and controlling of individual experiments.

Use a combination of agile, adaptive and eXtreme project management for the strategic planning, monitoring and controlling of a research project.

Find the quickest path to failure by doing the most difficult, highest risk and most important work as early as possible.

Incorporate data-based decision points, branch points and even pivot points at the end of each research project as part of your research program strategy.

8. **Make good decisions.** Research project management requires a prescribed series of decisions. The success of the research project and the utility of the knowledge it creates, depends upon the quality of those decisions.

Establish a culture within your R&D organization that accepts both individual and team-based decision-making.

Define the responsibility and authority of each individual on your team to make specific types of decisions.

Include key stakeholders in strategy decisions to facilitate acceptance and implementation.

Make decisions transparently for others to understand.

Be aware of the effort and time required to make consensus

decisions.

Enable individuals to make the decisions that affect the quality of their work.

9. **Embrace change and uncertainty.** Research project management requires a positive response to changes in assumptions, scientific models, hypotheses and external environment.

    Expect the unexpected and manage uncertainty through iterations, anticipation and adaptation.

    Recognize and mitigate risks. Monitor those with high risk exposure and react quickly to implement contingency plans.

    Recognize that some stakeholders have a low tolerance to risk and to change. These individuals require special communication.

    Recognize that variability in data contributes to uncertainty, obscures conclusions and requires confirmation beyond simply statistical analysis.

    Recognize that the decisions, branch points and pivots in your research project have implications on schedules, budgets and resource allocation. Your R&D organization must have the resource flexibility and a culture that effectively enables changes to tactical plans and schedules.

    Seize opportunities. Others will create change that is disruptive. Use creative approaches to turn change into opportunities.

    Enable flexibility. Ensure your team understands the objective clearly. Abandon an antiquated plan that is failing. Adopt an iterative workplan.

    Be willing (and able) to abandon the original plans and to adopt new tactical approaches or even completely new research strategies. Make pivot decisions if a better strategy arises to achieve the objective.

10. **Continuously improve.** Research project management requires that your research team learns from experience and improves both its scientific and management skills, continuously.

    Review your accomplishments and failures.

    Learn from each experiment, in terms of both science and management.

    Capture, communicate and implement the scientific and management lessons-learned at the end of each research project.

The Research Proposal

You can apply all of these principles as you conceive and conduct a research project. These principles of research project management form a set of process tools in the Scientist's Toolbox. The tools are applicable to all types of Observation, Modeling, Discovery and Development projects.

THE RESEARCH PROPOSAL

APPLICATION OF THE ten principles of research project management will help you to identify problems, visions and scientific models for your research projects. The following chapters will sequentially walk you through the process of creating a proposal that can be submitted to decision-makers and potential sponsors. Here are some points based on these principles to consider in this highly iterative, creative process.

1. **Stimulate creativity.** You will need creative and innovative solutions to a problem. Simply repeating what has already been done before or adding small incremental improvements is unlikely to gain support. I suggest a process that starts with divergent thinking to create lots of ideas, followed by convergent thinking to create a short list, then by consultation with others to obtain their feedback and finally by a selection decision.

2. **Satisfy customer's expectations.** Your research projects will have several customers who all want the knowledge from your research project for their own purposes. You must understand their needs and expectations. Their support will create a demand for your research project and help convince a sponsor to support it financially. Happy customers will promote your accomplishments, build your reputation and advance your scientific career. I suggest that you identify and focus on a primary customer but also provide your secondary customers with as many of their needs as possible.

3. **Build a strong research team.** Your research team must be committed. Together, they will face obstacles and will make important decisions as the experiments are being conducted. If they are involved in drafting the research concept, they will be more committed to achieving the objective and they will have a better perspective to make decisions. I suggest that the Principal Investigator involve members of the research team in conceiving a research proposal because this adds perspective. I suggest that you, as a member of the research team, actively participate in creating, reviewing and revising the research concept.

4. **Communicate effectively.** Financial support for your research proposal requires that sponsors understand the problem, the vision and the concept to achieve that vision. Clear communication in simple terms to decision-makers in a manner that provides understanding is essential. I suggest that you communicate the essence for your project in a three-sentence Research Project Skinny that all decision-makers can understand.

5. **Be Explicit.** Your research project will seek very specific knowledge about a scientific phenomenon. The concept must be explicitly stated as it is communicated to decision-makers. I suggest that you use a set of one-sentence statements that communicate the essence of the problem, your vision and your project's objective to allow everyone to understand explicitly.
6. **Keep it simple.** The complexity of scientific models, hypotheses and objectives can be overwhelming. Scientific models with the fewest number of assumptions are the best. I suggest that you propose a research project that can distinguish between two rival scientific models which make different predictions. You cannot disprove one scientific model without providing a better alternative explanation of the observations in the form of an alternative model.
7. **Plan carefully.** You need focus, clarity, critical thinking and logical arguments to create a rigorous research proposal that can withstand criticism from decision-makers. I suggest that you consider your research proposal from various perspectives. Consider the project from the perspective of different people, such as customers, stakeholders and decision-makers. But also consider the project from different scientific perspectives, such as ecological, biochemical and genetic.
8. **Make good decisions.** The quality of the selection decisions that you make in a research project will determine the quality of your results. You must select a problem, a vision of the future, a scientific model that explains relationships and hypotheses that can be tested. I suggest that you actively involve the research team, external "experts" or other interested people in the review of your proposal. Encourage a professional conflict of ideas from all perspectives, then decide. Not everyone needs to be involved in every decision. The best decisions are not always made by consensus.
9. **Embrace change and uncertainty.** Exploring the unknown is risky. There will be opportunities but there will also be dangerous pitfalls because the project has not been done before. When proposing a research project consider the pluses and minuses, the strengths and weaknesses, the opportunities and threats, the positive and negative forces affecting your vision and the scientific models that achieve your vision. I suggest that you identify and then focus on only the most impactful and the most probable uncertainties and develop contingency and mitigation plans.
10. **Continuously improve**. Writing a research proposal is an iterative process. I suggest that you avoid doing this alone. And avoid doing

this quickly. Take time, let your ideas incubate, plan, seek feedback and constantly improve the concept. Learn from your past experiences. Remember what you did well in your last project. Stop doing what did not work well and try something different this time. And most importantly, do not become discouraged, because there will be setbacks, just keep improving at each iteration.

KEY POINTS

RESEARCH PROJECTS ARE managed within some form of R&D organization that greatly influences the type of project management used.

Three important definitions are used throughout this book:
Research Project is a temporary endeavor with a defined beginning and end undertaken to create knowledge.
Research Program is a set of inter-related coordinated research projects undertaken to solve a problem.
Research project management is the management of talented, creative people to acquire knowledge.

Traditional project management practices are simply good management tools but they must be adapted to fit into the scientific method and the creative culture of a research team.

Research project management merges the concepts from traditional, agile, adaptive and eXtreme project management into the philosophy of the scientific method.

Traditional and research projects differ. Research project management is distinctive because:
- Knowledge is the Deliverable.
- Quality is Subjective.
- Conflict is inherent.
- Progress requires Decisions.
- Constraints Force Assumptions and Compromise.
- Uncertainty and Variability are Everywhere.
- Smaller is better.

Research project management is a management philosophy, not a management process, that facilitates communication and understanding. That philosophy is summarized in ten principles that are applied throughout the conception, proposal, conduct and closing of a research project.

Application of the ten principles of research project management will help you to identify problems, visions and scientific models as you propose a research project to sponsors.

PART THREE

CREATING IDEAS

CREATIVE THINKING

RESEARCH SCIENTISTS HAVE a common need to conceive ideas, to improve those ideas and then to select the best. Creative thinking has been highly correlated with the subsequently successful implementation of ideas[1]. Creative people have several common characteristics[2]. They think differently from others, but their creativity is not a birthright. Creative thinking must be learned, practiced, and nurtured. Creativity is first and foremost an attitude. In other words, if you think you are creative, you are likely to be creative. Even if you initially find it difficult, you can create new ideas by using tools to stimulate your creativity.

> *Henry Ford: whether you think you can, or whether you think you cannot, you are correct.*

Bloom's hierarchy of thinking[3] identifies creativity as the highest form of thinking. In this model, creativity is defined as the ability to conceive something new. The creative process requires five supporting skills as shown in the diagram below.

Creating	Synthesizing knowledge from different sources to create something new
Evaluating	Judging value, comparing ideas & identifying the strengths and weaknesses of facts, knowledge and scientific models
Analyzing	Understanding scientific models, reading scientific literature & seeing relationships
Applying	Using ideas & applying scientific knowledge to real situations
Understanding	Describing knowledge, summarizing & explaining scientific models
Remembering	Remembering the facts, terms, concepts & principles

Creative scientists use logic in the creative process to extrapolate from observations (empirical data), to make conclusions, to craft scientific models and to formulate hypotheses. Rarely do scientists have all of the facts, and what facts they have are often conflicting. Creative scientists, therefore, filter the data based on their experience, perspective, bias and intuition, when they create scientific models. They use their skill, craft and art to make assumptions, and then reorganize the filtered data based on those assumptions into scientific models using abductive logic[4]. Once those scientific models are in place, hypotheses that predict the outcome of experiments are proposed using deductive logic.

Creative scientists also use a combination of strategic, divergent, convergent and tactical thinking in the creative process. Strategic thinking articulates clear concepts and envisions opportunities. Divergent thinking emphasizes quantity more than quality, makes connections, seeks novelty and reserves judging until all possible options are on the table. In contrast, convergent thinking makes judgments, evaluates options and selects among alternatives. Finally, tactical thinking develops plans to achieve a specific goal.

In my Cycle of Innovation model, research projects address the four different categories of problems. The roles that you may play in your quest to solve those problems are different. I will call these scientists: Observer, Dreamer, Explorer and Builder, respectively. Observers prefer strategic thinking. They look at the big picture and see opportunities that others do not. Dreamers prefer divergent thinking. They are always coming up with ideas. Explorers prefer convergent thinking. They work through the options and the details to make objective selections. Builders prefer tactical thinking. They get things done.

We are all Observers, Dreamers, Explorers and Builders, but each of us has our own preferences and tendencies (e.g. FourSight[5]). Different people prefer and therefore tend to use different creative thinking styles[6]. Take a minute to do a self-evaluation. What type of creative thinker are you? Do you need to improve any of your creative thinking skills? Do you use explicit logical arguments, or do you use intuition? Do you prefer strategic, divergent, convergent or tactical thinking? Which do you least prefer? Your skills and preferences may influence the type of problems you address, the type of solutions you conceive and the type of research projects that you propose.

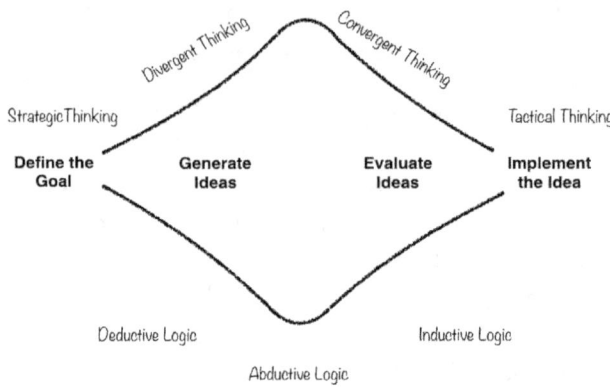

You have different preferences, different skills, different experiences and different intuition than your colleagues. When working in teams, your natural tendency is to work with people who think like you do. That tendency creates a collegial but limited team. The best research teams are diverse in knowledge, skills, perspectives and thinking preferences. The best research teams debate and embrace a conflict of ideas within a collegial environment.

Thinking isn't everything. Technical skill, craft and art are essential. Creativity requires hard work to generate, to evaluate and to implement ideas. Many ideas fail initially; perseverance is essential. Ideas and the scientific models on which they are based must be honed, polished, revised and improved before they can become the basis for great innovations. Once you have an idea for an innovation that will change the world, you are just beginning.

Thomas Edison: Genius is one percent inspiration and ninety-nine percent perspiration.

Different problems require different solutions. Close-ended problems have a single solution. For example, a mathematical problem has only one solution; or, there is only one way to fix a broken car. Most scientific problems, however, are open-ended problems that have many possible solutions, many different perspectives to consider and many alternative interpretations of observations. Fundamental to any research project is generating ideas - lots of ideas. Some ideas may just come to you like a lightning bolt, but this is unreliable. You might wait a long time for that lightning bolt to strike you. While waiting, you can use tools to look at the topic in different ways to generate and select ideas[7].

Several excellent websites are available with full descriptions of many creative thinking tools that may be relevant to your research projects, including American Society for Quality[8], Creating Minds[9], IdeaConnection[10], Litemind[11], MindTools[12] and Mycoted[13]. This section discusses some of the divergent thinking[14] tools and techniques to create ideas that were particularly relevant to my research projects and may help you.

Linus Pauling[15]: The best way to have a good idea is to have lots of ideas.

You need to create ideas both by yourself and with a team. Michalko[16], Silverstein, Samuel and DeCarlo [17]and other references in Further Reading describe numerous creative thinking approaches. These idea creation tools fall into several categories: free association, structured association, combination techniques, analogy and feedback. Free association techniques work well in group settings to create "out-of-the-box" ideas. Structured association techniques look at the problem from perspectives that are predefined by restrictive boundaries to improve an idea or find new applications for an existing idea. Combination techniques merge existing components from different sources to create something new. Analogy transfers established concepts from one scientific phenomenon to create ideas in another. Feedback solicits the considered opinion of others to improve ideas.

Edward De Bono[18]: Creativity involves breaking out of established patterns in order to look at things in a different way.

FREE ASSOCIATION TECHNIQUES

Free association techniques, commonly called brainstorming[19], gather a list of ideas spontaneously contributed by a group. This open creative thinking style is often needed in research projects that are exploring the unknown and doing something that has never been done before. Many variations have been implemented but they all collect as many options as possible and defer judgment on those options until later. This approach assumes that the odds of finding the best idea increase with the number of alternatives being considered. Brainstorming in an environment that is uninhibited, expanding and liberating is thought to produce more creative ideas compared to an inhibited environment with imposed restrictions on ideas or which only looks inward (internally within the team or within the R&D organization).

All brainstorming techniques have some common characteristics. When creating ideas, quantity is always more important than quality. Quick judgment may eliminate some potentially good ideas before you can consider all of the perspectives or iterations. Judgment inhibits creativity, in others and in yourself. Therefore, the process advocates that you retain your ideas and try to improve them, at least initially. Evaluation and selection will come later.

Tool # 1 *Brainstorming* creates a long list of diverse ideas and will be used many times as you conceive and then propose a research project. Your brainstorming sessions should be fun. The Principal Investigator may take the role of Facilitator for the meeting, but alternatively the Principal Investigator may delegate the facilitation of the session to avoid dominating the meeting and to avoid creating Groupthink[20]. The four basic steps in all brainstorming techniques as outlined in Tool # 1 *Brainstorming* are:
1. Gather all of the preliminary ideas.
2. Reorganize and compile the ideas into themes.
3. Discuss the options; compare their similarities and differences.
4. Add new options as they arise.

Criticism and debate are avoided at all times - this is difficult for scientists and requires special facilitation. Different approaches can be used to gather ideas. Ideas may be collected in a face-to-face meeting (Tool # 1.1 *Brainstorming Face-to-Face*), using cards (Tool # 1.2 *Brainstorming with Cards*), using structured tables (Tool # 1.3 *Brain writing 6-3-5*), using either physical or electronic bulletin boards (Tool # 1.4 *Brainstorming on Bulletin Boards*), or using a virtual team (Tool #1.5 *The Delphi Brainstorming Method*).

STRUCTURED ASSOCIATION TECHNIQUES

Structured association thinking techniques can be used to create, modify and improve ideas as part of a brainstorming approach or as a stand-alone process. Six Thinking Hats[21] explores the issue from different perspectives called hats. The hats have different colors. This is a collegial process where the collective team wears different "hats" sequentially to view an issue from different perspectives.

In the original concept, de Bono proposed that an issue be reviewed from the following perspectives:

White – Information: what are the facts?
Red – Emotions: what are the feelings?
Black – Caution: what are the risks? the negative aspects?
Yellow – Optimistic: what are the opportunities? the positive aspects?
Green – Creative: what are the new ideas?
Blue – Managing: what is the goal?

In a research project, you may want to use different "hats" than De Bono suggests. For example, in the life sciences, you may wish to view a topic from the molecular, biochemical, cellular, genetic, organismic and ecological points of view. A variation would be to view the topic from the perspective of different scientific models that make different predictions and hypotheses. Alternatively, you may wish to view a topic from the perspective of different customers or different stakeholders.

Note that Six Thinking Hats as described by De Bono, is a collegial, group activity. It is not a confrontational process; individuals do not advocate for one perspective versus the others. Instead, the group collectively views the issue from each perspective and only when all perspectives are understood, does the group make a selection or decision. This makes this approach exceedingly powerful for conceiving ideas in a research project when multiple scientific aspects or scientific models need to be considered.

Sometimes you need to create ideas on your own without a team. You may not want to convene a brainstorming session every time you need an idea. SCAMPER[22] is an alternative structured association technique that uses key stimulating words to create, modify and improve ideas. SCAMPER transforms an imperfect idea into a new better idea. SCAMPER analyzes the topic using seven different approaches. You can use one or more of these approaches. SCAMPER is an acronym for:

S – substitute; replace one attribute with another, e.g. color.
C – combine; merge two components to create a hybrid.
A – adapt; take a component from an analogous phenomenon.
M - modify or magnify; make a component very big or very small.

P - put to another use; find a new way to use something.

E - eliminate or elaborate; remove something; add something, e.g. a step in the process.

R - reverse or rearrange; turn a cause into an effect; advocate the opposite.

Both Six Thinking Hats and SCAMPER may be incorporated into a brainstorming session to stimulate idea creation. Remember to avoid making selections and to avoid criticism, because early judgement will inhibit idea creation.

COMBINATION TECHNIQUES

Combining existing elements from different sources will create a new something. Here are some approaches that you might use to stimulate idea creation either in a group brainstorming session or in yourself as you wait for your eureka moment. To begin, break a complex topic into smaller attributes[23], components or parts. The Morphological Matrix Technique[24] starts by listing these attributes as the column headings of a spreadsheet. Under each, list all of the possible characteristics of that attribute. For example, if two of the attribute columns are color and shape, the subsequent rows list all of the available colors in the one column and all of the available shapes in the second. To create a new something, one color is combined with one shape, either randomly or systematically.

A mindmap is an alternative way to list and rearrange the characteristics of different attributes showing relationships and dependencies. You can also use a checklist of the attributes in each of the topics to create a matrix diagram[25]. The Heuristic Ideation Technique[26] is another variation that has been used to create new products by combining elements of existing products.

ANALOGY

The premise of analogous reasoning[27] is that the solution to one problem may be used to solve another. Although this very efficient process is commonly used in scientific research, it can be misleading if the cause-effect relationships are poorly understood or different in the two phenomena. Two forms of analogy are commonly used.

The first is to borrow ideas for other scientific phenomena. For example, to create new hypotheses about the treatment of pancreatic cancer, you might list the known treatments of other types of cancer: breast, colon, lung, brain, liver, etc. Are there any treatments of the other types of cancer, that might be hypothetical treatments for pancreatic cancer? As another example, to create ideas on how to improve the yield of wheat by plant breeding, you can examine the methods used in maize breeding. Are there any breeding approaches that have been successfully used in maize that might be hypothetical ways to improve yield in wheat?

The second is to mimic what nature does. Biomimicry[28] is a process that adapts nature's solutions to solve our problems. The Wright brothers studied the flight of birds and then used that knowledge to design their first flyer[29]. Another example of biomimicry is George de Mestral's observations of burrs that were stuck to his clothing[30]. He examined them under a microscope and noted that the cocklebur has thousands of tiny hooks that attach its seeds to fur and feathers. He copied this to create a system of hooks and loops in Velcro that when pressed together stick to one another but separate easily. The Biomimicry Institute[31] follows "an approach to innovation that seeks sustainable solutions to human challenges by emulating nature's time-tested patterns and strategies". Their website gives many examples of biomimicry in products that we use every day.

FEEDBACK

Just as you will rarely have a eureka moment, you will rarely create a scientific idea completely from scratch. Great inventors work with others collegially. They take ideas from others, modify them, add their own perspective and create something new. Jain, Triandis and Weick[32] expressed the opinion that 75% of the creative ideas originate from ideas that were communicated verbally from other scientists and only 15% come from individuals reading the scientific literature.

Improve your ideas by actively requesting feedback from others. Feedback[33] can identify strengths and weaknesses in your ideas, allowing you to modify and to improve them. When seeking feedback, it is essential that you communicate effectively, and that the reviewer fully understands your idea or proposal. Negative feedback is often based on misunderstandings or poor communication. Know what you want to learn and ask the right questions. The better your questions are - the better the feedback. Be specific in your questions. Avoid asking open ended vague questions. Seek feedback from as many scientists and from as many perspectives as feasible. You can informally ask individuals in a conversation or formally ask a panel of experts (Tool # 3.1 *Expert Panel*), a virtual team (Tool # 3.2 *The Delphi Method*) or interested stakeholders (Tool # 3.3 *Deliberative Forum*).

You can guide your reviewers through the feedback process by asking leading questions. DeLong [34] advises using a feedback mechanism called SKS that asks for advice from others on what you should Stop, Keep and Start. Specifically, he advises to ask three questions:

- What should I stop? or What should I remove?
- What should I keep?
- What should I start? or What should I add?

Many of the tools in The Scientist's Toolbox include a step to solicit feedback. A short explicit statement that is easily understood by your reviewers is a common output from many of these tools.

Unfortunately, negative feedback is common in science. Providing negative feedback to others may not achieve what you intended. Buckingham and Goodall[35] point out that criticism based on your perception of someone else's mistakes or shortcomings may be quite counterproductive. They advocate giving positive feedback that encourages others to build upon their good performance and their creative ideas[36]. To avoid being overly critical and judgmental, consider using POINt to provide feedback on the ideas and proposals of others (Gerard Puccio[37]). POINt is an acronym for Pluses, Opportunities, Issues, New thinking. When providing feedback on an idea you should sequentially:

P: List the **positive** aspects of the idea, all of the pluses. All ideas have good points and it is best to discuss these first. This gives positive reinforcement to the person proposing this idea. People generally tend to be defensive about their ideas. If you present the positive aspects first, this defensive reaction is minimized and makes people much more receptive to your suggestions and feedback.

O: List all of the **opportunitie**s that the idea may present. What benefits might be achieved? What might we do if this idea works? Who will benefit?

I: Raise the **issues** that you see with the idea. What are the challenges?

What could go wrong? What are the negative aspects? These are the criticisms.

NT: Indicate where **new thinking** is needed to improve the idea, to create new ideas and concepts around each of the issues and to reformulate the idea. How can this idea be improved? How can the negative issues be resolved? This ends the feedback with constructive ways to improve the idea.

Many scientists lack skill in providing feedback. You should anticipate negative and sometimes unflattering, non-constructive comments. When you receive negative feedback, you should accept the comments as given. Avoid debating, responding defensively or making excuses. Request specific details and examples to ensure that you understand the concerns and comments. You should thank the messenger for an honest opinion, even if this is difficult, but then you should verify the feedback with others. If one person has concerns, likely others have the same unexpressed concerns.

The action that you should take after receiving negative feedback depends upon the reasons why your colleague had a negative opinion. Once you understand the underlying reason, you can take corrective actions.

- Misunderstanding based on communication failures and inadequate information.
- Disagreement about the facts, scientific models, scientific assumptions and hypotheses.
- Disagreement about the logical arguments that led to the conclusions made.
- Disagreement about the quality of the data or information used.
- Intolerance to change, lack of job security or lack of organizational stability.
- Self-interest because the idea may threaten an individual's own interests.

CONVERGENT THINKING

At the end of a brainstorming session, you will have a long list of ideas that require follow-up actions. Brainstorming is only the beginning. Burkus[38] contends that to produce innovative breakthroughs you need to do more than throw ideas around. Hard work is needed to develop the off-the-wall ideas from brainstorming into the reality of new scientific knowledge and technology.

A convergent thinking style is needed to sort through options critically and to pick the best one for your research project. Kahneman[39] observed that people tend to use two different approaches to make decisions. The fast one, called System 1, is an intuitive and emotional approach using experience and beliefs. The slow one, called System 2, is a deliberate, rational and logical approach requiring detailed, methodical consideration. People have a tendency to use System 1 to make decisions quickly, and jump to conclusions, even for complex problems, and scientists are no different. The decisions made by fast thinking just feel right. Kahneman says that this reflects an overconfidence in our abilities and the mental laziness of System 2. For a research scientist, fast selection decisions are usually poor decisions. Fast thinking inhibits interaction with others. Fast thinking disrupts a research team. I am advocating in research project management that you follow a slow methodical process to make project selection decisions. There are many decision-making guides[40] and references to aid you in making decisions - just type in a Google search. Although it may initially seem to be laborious and tedious in a research project, the transparency and quality of your joint decision-making will justify the time and effort.

SELECTION CRITERIA

YOUR SLOW, METHODICAL process to make a selection decision involves several steps. Tool # 2 *Categorical Selections* is such a multi-step process that you can use in many different situations during a research project. You can use one, two or more of the steps depending on the complexity of the selection. The objective of your selection process is to make one of three possible decisions about the options.

A: Select options that pass a threshold to qualify.

B: Rank the options from high to low.

C: Select the best or "winner takes all".

Your first step is to quickly shorten a long list of options using MUST HAVE criteria. You often use Tool # 2.1 *Pass-Fail Criteria* in your daily life to discard options. For example, if you are buying a house or a car, you probably use a set of MUST HAVE criteria. You can create the criteria yourself or brainstorm a list with your team. You can make the process transparent to gain acceptance from stakeholders by sharing the criteria or jointly developing the criteria. You can use your intuition, your experience and your background knowledge as you rate ideas into YES and NO categories for each criterion. Then discard any option that fails to meet all of these criteria. The pass-fail analysis either shortens a list of options for further analysis or identifies those that pass a threshold (Type A selections).

Pareto Analysis[1] applies the 80:20 rule (Tool # 2.3 *Pareto Analysis*) as an alternative approach for Type A selections (See Koch for more details[2]). The Pareto Principle states that roughly 80% of the effects come from 20% of the causes. For example, the rule states that 80% of the wealth of a society is held by 20% of its population. The Pareto analysis identifies the options that have the largest effect enabling you to focus on those ideas that have the greatest impact or value[3]. This analysis has particular value in risk and safety assessment, but it can also be applied to hypothesis and strategy selection.

As a simple example, consider the following biochemical pathway: A > B > C > D. Many factors may influence this biochemical pathway, but they will have a different quantitative stimulation of the flow through the pathway. The degree of stimulation that each factor has on the pathway can be quantified and ranked from high to low effects. Those factors that have a cumulative effect equal to 80% of the total stimulation become the focus for further consideration. The factors that contribute the remaining 20% of the effect are discounted[4].

There are several alternative second steps depending on the type of selection decision that you want to make. An all-purpose method of subjective selection is PMI Analysis[5] that addresses the Pluses, the Minuses and the Interesting aspects of an idea. This enhanced pros vs cons analysis was proposed by Edward De Bono[6] as a method to think laterally. This method encourages you to evaluate an idea from more than one perspective and to refine an idea by focusing on its interesting aspects. This process can be adapted to rank options (Type B selections).

A relatively simple process that I have included in Tool # 2 uses a set of non-essential but nice-to-have positive criteria that collectively can be called Wants. These represent opportunities or benefits of an option. These criteria are rated categorically as Yes/No (Tool # 2.2 *Wants and Limitations Criteria*). Next, consider the negative aspects of an option that can collectively be called Limitations. These criteria represent potential constraints or adverse effects, threats and risks that should be avoided. These criteria are rated categorically as Yes/No (Tool # 2.2 *Wants and Limitations Criteria*). The Wants and Limitations ratings can be done by either an individual or a group.

Subjective analysis, giving categorical assessment as Yes or No, relies upon your personal beliefs. In some situations, this may be an advantage, but in others, it is a distinct disadvantage, especially if your selection will be scrutinized or criticized as being biased. If your decisions are based solely on your subjective personal analysis, you will find that it is difficult to justify them to managers, to stakeholders or to investors.

Several objective analysis tools remove bias (or at least make your biases more transparent). The collective ranking by several people is perceived as being more objective than one done by an individual. A more quantitative assessment is the decision matrix[7] (Tool # 2.4 *Decision Matrix*) that makes a numerical rating (perhaps on a scale of 1 to 10). The matrix is a list of criteria and values in table format that allows you to identify, analyze, and rate them. For example, if you are buying a car, the criteria might be cost, fuel economy, cargo space, safety, comfort and attractiveness. The first four are quantitative criteria that be obtained from the car's spec sheet, but the last two measure your subjective opinion. To make the selection even more stringent, you can assign differing importance and consequently different weighting to the criteria when calculating a final score. For example, the cost of the car might be more important than its attractiveness. This decision matrix will provide a ranking for each criterion and a total score for each option, just like the Consumer Reports ranking of cars. Therefore, this process can be used for Type B selections.

Pairwise Comparison Analysis[8] (PCA) uses pairwise comparisons sequentially for a long list of options that are relatively simple to distinguish (Tool # 2.5 *Pairwise Comparisons Analysis*). The method is useful to integrate several criteria, when the criteria and their priority are ambiguous, when their assessment is subjective, and when the options are completely different. PCA method works well for relatively simple Type C selections.

Analytic Hierarchy Process[9] is an enhanced, quantitative variation of PCA that can be used for Type C selections when multiple, potentially conflicting criteria need to be considered in each comparison. Instead of evaluating options as a whole entity, the AHP dissects the PCA selection into want and limitation criteria that can be analyzed independently. Tool # 2.5 *Pairwise Comparisons Analysis* can be adapted to use AHP by repeating the pairwise comparison for each criterion and then summing the total score.

REVIEW PANELS

ONCE IDEAS PASS a preliminary selection process, such as Tool # 2.1 *Pass-Fail Criteria*, a more detailed technical evaluation may be warranted before a final selection decision is made. Research projects are highly technical and ideas for research projects, hypotheses, experimental designs and research strategies require detailed technical evaluation, perhaps by a panel of technical experts or by a panel of interested stakeholders.

Some R&D organizations use an adversarial review panel that passes recommendations to senior management, not to the research team. This confrontational process rarely improves the quality of the research proposal. It simply discourages creativity and stifles innovation before concepts can be fully developed. Instead, my view of a review panel is one that is charged with a mandate to explore alternatives and to advise the research team in way that improves the impact and probability of success of a potential research project, not a judgement.

I describe three forms of a review panel in Tool # 3 *Review Panel*. The structured discussion among experts to make recommendations on a specific topic is what I am calling an expert panel. This analysis is conducted to gain a deeper understanding of the state-of-the-art and the participants' informed opinions concerning the topic. However, because of the huge time commitment, use expert panels only for the most critical topics concerning a scientific phenomenon that are technically complex. Tool # 3.1 *Expert Panel* is an intensive and expensive way to obtain feedback on ideas from experts that can meet face-to-face.

An alternative version is described in Tool # 3.2 *The Delphi Method* to obtain feedback from geographically dispersed experts. The Delphi Method [1] is an iterative process that uses a series of questionnaires to generate written discussion and argument.

The participants in a review panel interact actively but the organization is flexible. The specific format depends whether the topic is being reviewed by a team that meets face-to-face or by a group of experts scattered across multiple locations. Usually, the expert panel is a one-time meeting, but in some instances several meetings may be required if the topic is highly technical with iterative information gathering, sharing and discussion. If you are using Tool # 3.2 *The Delphi Selection Method*, this meeting may span several weeks.

You should select a group of experts based on their diversity of technical knowledge, opinions and perspectives relevant to the topic.

The participants usually include both internal and external (e.g. consultants) technical experts. The background should be shared in advance to allow the participants time to prepare. The meeting should focus on a specific question. Avoid using questions such as "How might someone do …..?" or "How might X cause Y?" that lead to brainstorming. Avoid open-ended questions such as "How are we doing?" or "What do you suggest?" Instead, ask a specific technical question, such as "Which of these options is the best way to achieve ….. ?"

At the meeting, you should encourage all participants to discuss the topic until all facts and technical details are collected. The scientific merit of the model and hypotheses may be actively discussed. Assumptions inherent in the scientific model may be critically evaluated and challenged. Unlike a brainstorming session, you should encourage the participants to advocate for their preference and freely criticize other options. In other words, debate is part of the creative process. You should encourage and facilitate this debate, but at all times it must remain professional. This is difficult because opinions are often based on core beliefs. When anyone's core beliefs are challenged, the conflict quickly becomes personal.

The expert panel should not be empowered to make decisions. The panel should only advise, provide feedback and recommend rankings. The panel lacks accountability and responsibility. Finkelstein[2] raised several other concerns about decisions made by experts, especially in science-based industries. An expert in a scientific discipline became an expert by either developing or validating the scientific paradigms in their discipline. Their experience and knowledge is based on empirical data, has solved problems in the past and has explained a scientific phenomenon to their peer's satisfaction. Kuhn[3] observed that the expert has a strong preference for the current paradigms and is reluctant to change. Because the expert knows the scientific discipline and the current paradigms so well, the expert may cease to be curious. The expert falls into what Finkelstein calls the "expertise trap". Scientists are particularly vulnerable. These trapped experts are unlikely to support challenges to scientific paradigms. The experts' overconfidence darkens their perspective of a new innovation. These experts may support an adaptive innovation that offers refinements to the current scientific paradigms, but they will oppose radical innovation.

Colin Powell[4]: Don't be buffaloed by experts and elites. Experts often possess more data than judgement.

Your expert panel may lack a diversity of opinion, because by definition they are experts. The participants in your expert panel are likely very busy and may devote little time to think about your radical ideas that are "out-of-the-box". In these cases, consider using Tool # 3.3 *Deliberative Forum*. A deliberative dialogue[5] is a moderated process for a group to discuss political and social issues and is a very effective way for citizens to become informed. The process begins with the preparation of an issue guide that outlines a contentious topic and presents alternative solutions for the group to discuss. After reading the guide, the participants express their initial concerns and opinions. Sometimes, the discussions of social and political issues can become passionate because the topic impacts an individual's core beliefs. The discussion highlights these tensions, explores concerns and identifies areas in which people are resistant to change. New ideas are often expressed. The dialogue does not make a selection decision but provides guidance and insight. This may be in the form of feedback from interested stakeholders about your ideas, identification of tensions that may lead to resistance and identification of areas of agreement that might lead to support. Tool # 3.3 *Deliberative Forum* is a modification of this process that is applicable to a research project as an alternative or as a supplement to Tool # 3.1 *Expert Panel*. Note that this is a time-consuming process and there are four critical success factors to consider.

1. Prepare an informative, unbiased guide that outlines options with both positive and negative arguments. This guide is critical to frame the discussion.
2. Select your participants with a diversity of expertise, experience, thinking style and beliefs, but avoid the experts; they should be advocates, but not argumentative; they should be creative, but pragmatic; they should be collegial, but think independently.
3. Moderate the discussion actively and proactively. This is a demanding role that requires skill and training.
4. Ensure that the feelings, concerns and biases of each participant are transparent.

GROUP DECISION-MAKING

AT THE END of an analysis or review using either Tool # 2 *Categorical Selections* and/or Tool # 3 *Review Panel*, you will have a prioritized list of options. A final selection decision must be made by the research team. Decisions do not just happen; they must be managed. The selection process must be transparent and accepted by all participants and stakeholders. Usually this involves some form of voting. Have you ever participated in a long discussion that ended by the facilitator saying, "I think everyone agrees, so that is what we will do"? This may have saved valuable meeting time, but did everyone actively support the decision? Did everyone leave the meeting with enthusiasm? Maybe. The extroverted participants who dominated the conversation may be pleased. The introverted, quiet participants may feel excluded.

Instead, I recommend that you use an explicit voting process to make the final selection decision (Tool # 4 *Group Voting*). Your voting process should consider the type of selection decision needed, its importance and the need for group support to implement the decision. The more important the decision is to the project, the more people should support it. Tool # 4.1 *Multivoting* will create a prioritized list of options giving everyone an equal voice in the ranking. If the team will accept any selection without opposition, then a majority voting process is the simplest way to make a decision. If you want to have support without anyone resisting, then consensus voting (Tool # 4.2 *Consensus*) is required. If the enthusiastic support of everyone is required, then the choice should be unanimous.

Consensus[1] has been promoted, discussed and analyzed in most project management manuals. It is also the most misused decision-making process. I will only give a few highlights and reminders of the process that seem relevant to research projects. If your group decides by consensus, it does not mean that the decision is unanimous; it means that everyone can more-or-less support the decision. However, more-or-less[2] is a subjective, ambiguous term.

To reach consensus, the facilitator of the meeting explicitly asks all participants if each can more-or-less support the first choice. If everyone can, that is the selection decision. If anyone is opposed to this choice, the process is repeated for the second option. This continues until everyone can more-or-less support one of the options. If some level of support for any option is impossible, or if there is no option to which someone is not opposed, this fact is included in the meeting report with reasons.

Someone will then make an autocratic decision. Even if no one makes a decision, that inaction in a research project is in effect a decision to do nothing.

The consensus decision-making style is commonly thought to reach much better decisions than decisions made by individuals. This is because the team can pool its collective knowledge, perspectives, biases and individual beliefs, thereby merging its collective experience and wisdom. At least, that is the theory given in most project management and team building manuals.

> *Abba Eban: A consensus means that everyone agrees to say collectively what no one believes individually.*

If a decision is made by consensus, most people feel that it is the right decision. This is a fallacy because several pitfalls and problems with consensus decisions go unrecognized by most teams. A major problem is that some participants are reluctant to raise objections or appear as a "red arrow". Instead, they feel that collegiality is more important than critical decision-making. Therefore, they will support a decision, even if they think there is a better option, without voicing their opinion.

> *James Lewis[3]: The false consensus effect is a failure to manage disagreement, because no one is willing to express disagreement.*

Some Principal Investigators are reluctant to debate and actively promote a false sense of collegiality by inhibiting debate within their team. Some Principal Investigators feel that silence is agreement. Some Principal Investigators dislike being challenged. Some Principal Investigators cannot manage creative professional conflict. These Principal Investigators quickly assume consensus without actively seeking it.

Another potential problem with consensus decision-making is the most common and the most difficult to detect and correct - Groupthink. This pitfall was recognized first in the business community. Irving Janis[4] originally proposed that Groupthink occurs within a group of people who have:

- A strong desire for harmony.

- An aversion to conflict.
- Worked together previously on successful projects.

As a result, the group has overconfidence in the quality of its knowledge and decisions. They fail to critically evaluate alternative options, risks or threats; they suppress dissenting viewpoints, even their own; and they isolate themselves from outside influences. They begin to make poor choices because Groupthink inhibits individual creativity, uniqueness and independent thinking. Groups that develop Groupthink make decisions like a biased individual. They miss opportunities and their projects eventually fail.

> *Elizabeth Warren: Groupthink can become a serious issue - old ideas stay around after they're useful and new ideas too often don't get a fair hearing.*

Groupthink in R&D organizations always starts with good intentions. Sponsors have scarce resources and R&D organizations compete for those scarce resources. Any criticism or dissension may jeopardize funding, even if it is constructive, well-intentioned criticism. Managers in R&D organizations emphasize harmony and avoid disagreement; they exert strong pressure on individuals to conform and present a unified front. R&D organizations create reward systems that recognize collegiality and teamwork but fail to balance collegiality with critical thinking.

When an R&D organization suppresses disagreement and restricts dialog, individuals act in a restrained manner. Most people wish to avoid being marginalized, losing influence or being excluded from a team. They fear being considered an "ineffective" team player by their peers. As a result, the individual self-censors and avoids presenting dissenting data, perspectives or opinions. This behavior becomes part of the R&D organization's culture[5]. The project management manuals in Further Reading suggest several techniques that leaders of a team-based decision-making can use to avoid Groupthink.

In summary, I advocate that you use Tool # 1 *Brainstorming*, Tool # 2 *Categorical Selections* and Tool # 3 *Review Panel* sessions sequentially in a gated process incorporating Tool # 4 *Group Voting* at each step to select the best ideas. You should keep the steps distinct. A process that separates idea creation from a technical scientific debate ensures that ideas are fully vetted and prioritized. Selecting only the best ideas for expert review saves time and scarce resources for more important tasks. This multi-step process will present your team selection decisions in a transparent manner to managers, customers and potential sponsors.

KEY POINTS

WHILE WAITING FOR a lightning bolt of inspiration, you can use divergent thinking tools to collect lots of ideas. Ideas should be collected without criticism or judgement because that inhibits creativity in both others and in yourself.

Tool # 1 *Brainstorming* can be used in different formats to stimulate idea creation. The distinction is that the participants should be fully briefed on the technical details. Although less spontaneous, this is a prerequisite for brainstorming on a scientific topic.

Alternative approaches and tools that I found useful to create ideas for research projects include:

- Explore the issue from different perspectives using Six Thinking Hats.
- Transform an existing something into a new something using SCAMPER.
- Break a complex topic into smaller attributes, components or parts using a Morphological Matrix.
- Combine elements of existing ideas in a new way using the Heuristic Ideation Technique.
- Borrow ideas for other scientific phenomena using analogous reasoning.
- Mimic nature using biomimicry.
- Improve your ideas by requesting feedback from others.
- Request and provide positive feedback to others using POINt.

There are three different categories of selection decisions that are needed often in research projects:

A: Pass a threshold to qualify.

B: Rank from high to low.

C: Best, "winner takes all".

Tool # 2 *Categorical Selections* is a convergent thinking tool that selects the best ideas using a multi-step process.

To shorten a long list of options and to make Type A selections, use a set of MUST HAVE criteria as described in Tool # 2.1 *Pass-Fail Criteria* or use Tool # 2.3 *Pareto Analysis*.

To prioritize options and to make Type B selections, use subjective analysis methods that reflect your experience, informed judgment, intuition and opinions. Subjective scoring tools include PMI Analysis and Tool # 2.2 *Wants and Limitations Criteria*.

To prioritize options and to make Type B selections, OR to make a winner selection and to make Type C selections in a manner that reduces bias and increases transparency, use objective scoring involving numerical evaluation of multiple criteria. Objective scoring tools include Tool # 2.4 *Decision Matrix*. and Tool # 2.5 *Pairwise Comparison Analysis*. Analytical Hierarchy Analysis uses repetition of Tool # 2.5 *Pairwise Comparison Analysis* for multiple criteria.

To obtain recommendations on a specific technical topic, use Tool # 3.1 *Expert Panel* in a structured discussion among technical experts in a face-to-face meeting. You will gain a deeper understanding of the state-of-the-art and the participants' informed opinions concerning the scientific phenomenon. Use Tool # 3.2 *The Delphi Method* if the experts cannot meet face-to-face. Beware, both are intensive and expensive ways to obtain feedback.

Tool # 3.3 *Deliberative Forum* is a modification of a deliberative dialogue, which is a moderated process for a group to discuss contentious issues. The feedback identifies tensions and support for ideas.

Your choice of a group voting method to make a selection decision using Tool # 4 *Group Voting* depends on:

The type of selection needed.

The importance of the selection.

The need for group support to implement the decision.

Consensus decision-making means that everyone can more-or-less support the decision, even if it is not everyone's first choice. Beware, Groupthink in consensus decision-making inhibits creativity, uniqueness and independent thinking. If it occurs, the group will eventually miss opportunities and their research projects will fail.

PART FOUR

FINDING SOLUTIONS TO PROBLEMS

THE PROBLEM

RESEARCH PROJECTS CREATE knowledge in an attempt to solve a problem. The problems are diverse. Different people have different problems. Some view a problem in project management terminology as the gap between where they are and where they want to be. Others view a problem in academic terms as a puzzle that needs to be solved. Some want to improve peoples' lives, cure a disease or prevent starvation. Others want to capture a business opportunity. Still others want to understand our world better. All seek to acquire more complete knowledge about a scientific phenomenon. The lack of knowledge is the common problem that motivates scientists and justifies any research project.

Different problems require different research projects. In research project management, I found it useful to distinguish among research projects based on how they seek to address four different categories of problems.

A. Problems that are poorly described require an Observation project to observe events, to measure effects, to collect more facts or to convert qualitative into quantitative measurements.

B. Problems that lack understanding require a Modeling project to learn about relationships, to predict outcomes or to improve the scientific model.

C. Problems that have several hypothetical solutions require a Discovery project to select the "best" solution or to test the validity of a hypothesis (prediction) in different situations, and in the process, they may discover a previously unknown relationship or invent something new.

D. Problems that need validation of a solution require a Development project to produce an innovative technology or a prototype solution based on the prediction of a scientific model.

Albert Einstein: If I had an hour to solve a problem, I'd spend 55 minutes thinking about the problem and 5 minutes thinking about the solution.

The selection of which problem to address in research is the most significant decision that a scientist can make. Everyone wants their research to address an important problem. Importance is, however, a vague and subjective term that may mean financial, scientific, societal or personal importance. Importance may be measured by the number of people affected, by the magnitude of the effect or by the anticipated financial gain. Therefore, the assessment of importance is your personal subjective assessment.

Many national and international agencies describe problems that require more scientific knowledge to solve. For example, the United Nations in 2000 established eight Millennium Development Goals[1] for improving the lives of the world's poorest people, which guided activities of the United Nations system at the country level. The Bill and Melinda Gates Foundation fosters innovation to solve global health and development problems by funding Grand Challenges[2]. Many agencies, such as the World Economic Forum[3], conduct surveys to identify people's wants.

You may assume that only applied or commercial research projects address problems. However, academic research also addresses scientific problems that need to be solved. In this case, a scientific model may have gaps, inconsistencies or inadequately explain observations. A poorly defined scientific model is a problem. An untested hypothesis is a problem. An assumption is a problem.

How a problem is framed provides its focus and emphasis. The problem is often phrased as a loss or the negative consequence of something. Consider the following, related but different, wordings of a problem (National Cancer Institute[4]):
1. More than 40,000 women die from breast cancer every year.
2. More than 250,000 women are diagnosed with breast cancer every year.
3. More than 10% of the women diagnosed with breast cancer die within 5 years of diagnosis.

Each wording expresses a loss, but in different terms. The different terms suggest a different aspect and a different focus for research on breast cancer. Your perspective and your bias shape the problem that you will address. Instead of wording the problem as a loss or as a negative consequence, you could frame the problem in a way that allows you to improve or gain something. How the problem is framed may allow you to envision different solutions. For example, consider the solutions that you might envision from these wordings of the problem:
A. We need to reduce the number of deaths from breast cancer.
B. We need to reduce the incidence of breast cancer.

C. We need to extend the survival of patients after a diagnosis of breast cancer.

> *Bertrand Russell: The greatest challenge to any thinker is stating the problem in a way that will allow a solution.*

Therefore, think about how you frame and word your problem carefully. This may be an iterative process as you gain more understanding. You may frame the problem differently for different customers or for different sponsors. A problem that is stated clearly and succinctly in a manner allows others to understand it, to set a focus and to enable a solution.

> *Charles Kettering[5]: A problem well-stated is half-solved.*

A literature review is the most common way to understand a scientific phenomenon. Jain, Triandis and Weick[6] observed that the most creative researchers have a good general knowledge of the topic, understand the current scientific models and have the freedom to create new concepts. You can strive to learn more about the problem by using some of the approaches detailed in Tool # 5 *The Problem Statement*. The project management manuals advise you to dig deeper into a complex topic using the repeated questions technique[7]. By repeatedly asking either "Why?" or "How?" this variable is causing this problem, the root cause of a problem may be identified. By adding " ... else?" to the question it is possible to think laterally and to add breadth. You can follow this exercise by asking "So what?" to determine the consequences and relative importance of different potential causes of an effect.

When studying a new problem, maintain a broad perspective, so you consider all options and potential causes of a problem. Do not get mired in detail or fall into the expertise trap[8]. Do not rush to find a solution. More scientific detail will be added later. A summary of the problem can be written in a Problem Description for technical review and future reference. The main conclusion of this report should be why this problem is important to solve.

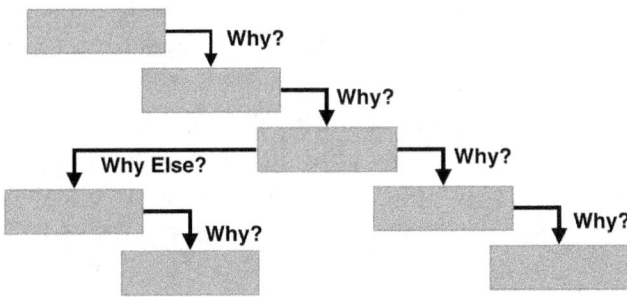

Albert Einstein[9]: The mere formulation of a problem is far more essential than its solution, which may be merely a matter of mathematical or experimental skills. To raise new questions, new possibilities, to regard old problems from a new angle requires creative imagination and marks real advances in science.

Writing a good Problem Description is an iterative process. You will normally begin by writing a draft that details the information that you have collected. As you dig deeper and understand the topic more clearly, you will revise the Problem Description. Your description is a critical, interpretive assessment of current knowledge. A potential customer at this stage is simply a best guess at a group who will use the knowledge for their professional or commercial purposes. Based on your Problem Description, the next step is to write a one sentence Problem Statement:

[WHO] needs [WHAT] because [WHY].

The advantage of stating the problem in one sentence is that the brevity provides clarity. An explicit Problem Statement provides focus. A simple and clear Problem Statement ensures that everyone understands.

Taking the above example of breast cancer, all of these problems are important to solve. The solution of each will provide enormous benefit. But realistically, you cannot do everything in one research project. What specific problem are your trying to solve? Your analysis of the problem and review of the literature may result in the following wording for a Problem Statement:

Medical doctors need a treatment to reduce the incidence of breast cancer because more than 250,000 women are diagnosed with breast cancer every year.

Note that this statement could have been worded in many different ways addressing different aspects of breast cancer. Recognize that its wording skews your perspective in different directions leading to different visions, different objectives and different concepts for research projects.

What scientific problems are you considering? Tool # 5 *The Problem Statement* provides you with step by step guidance on how to analyze a problem to determine if it merits your attention, your time and your skill. If you are going to start a research project, you are going to be committing a lot of your time and a lot of money. This analysis is crucial. This decision will define your career.

As an exercise, write a Problem Statement for your current research project: *[WHO] needs [WHAT] because [WHY]*. Change your wording of the Problem Statement. Do a thought experiment to determine the effect of different wording on your expectations. Which is the most inspirational? You might also ask others on your research team to write their Problem Statement. Do this independently, without cross referencing. If you find that their Problem Statements are different from yours, your colleagues are working towards solving a different problem than you are. They will have different motivations and expectations and this might create misunderstandings and conflict in the future.

The Problem Statement is a concise way for you to align expectations, to communicate, to justify your proposal to customers and sponsors and to motivate others, as you recruit your research team. The more detailed analysis written in a Problem Description will be included in legal, business and scientific reviews, and in grant applications.

OBSERVATION PROJECTS

OBSERVATION PROJECTS ARE conducted to study problems that are poorly described. Initially, the exact problem may be obscure. The scientific phenomenon may be vaguely defined. The fundamental scientific question may be illusive. Contradictions may exist in the published data. Critical observations may be missing. The scientific model may not suggest a testable hypothesis. Many fundamental relationships are unproven. A more precise definition of the problem may determine its cause and suggest potential solutions.

Even though people have wants and needs, they may express them poorly. They may know there is a problem but not know the cause. They may wish to take advantage of an opportunity but do not know how. Many commercial R&D organizations spend considerable time and effort to understand problems that might become business opportunities. The intellectual property rights may be obscure. The commercial opportunities may be uncertain. The market may have a long diffuse value chain. The potential financial returns may be unknown. Analysis of intellectual property, markets and value chains helps the business to determine what products to develop and how to capture value from them.

Henry Ford[1]: If I'd asked people what they wanted, they would have said faster horses.

The Observer seeks to understand the current situation, to create a unique perspective and to envision what the future may hold. The Observer collects facts, catalogs information and searches for problem-solving opportunities. The Observer seeks to understand the relationships that cause the problem. In the Observer's quest to describe a scientific phenomenon, the Observer delves into the root causes of a problem. The Observer selects the variables to measure and how to measure them based on prior experience.

An Observation project collects data with an experimentalism philosophy[2]; it does not propose and test explicit hypotheses. Instead, the Observer collects data that describes a scientific phenomenon. The data may be qualitative, quantitative or analytical. The data may include first hand observations of the phenomenon. The data may include quantitative measurements of a phenomenon, such as the changes that occur during the development of a disease. The data may be financial or economic data that are analyzed to convince a business leader or investor that further research will generate a financial return.

A generic Observation project has the following tasks in its workplan:
1. Conduct a preliminary review of the scientific literature to understand the background knowledge on a scientific phenomenon or a problem.
2. Summarize the problem using different words or different perspectives.
3. Divide and fractionate the problem into simpler components.
4. Delve deep by asking repeated questions such as Why? or How? or So What?
5. Think laterally by asking Why else? or How else?
6. Gather facts; observe the problem first hand; make measurements; collect data.
7. Understand the relationships and interactions among the components of a problem.
8. Identify the root cause of the problem; propose cause-effect relationships.
9. Challenge the assumptions in the current scientific models.
10. Communicate a story.

Most research scientists play the role of an Observer in their research projects that collect observations and empirical data about a scientific phenomenon. Charles Darwin was an Observer when he collected biological samples during his voyage on the HMS Beagle. Gregor Mendel was an Observer when he classified and counted his pea seeds. Their observations led to fundamental scientific models for evolution and genetics, but those scientific models were based on the empirical data they collected as observations. Observation projects may easily, and quickly, transform into Modeling, Discovery or Development projects when new insights are gained that suggest new scientific models, discoveries or innovations.

The deliverable of an Observation project is a communication that describes some aspect of a scientific phenomenon. The communication is written either as a scientific publication or a confidential analysis of a business opportunity. The communication may also be in the form of a data archive or a searchable database, which is the most common depository for large datasets, such as genomics data[3]. The customers are often other scientists who are researchers, advisors or teachers who apply the information, but the customer may also be a business leader or an investor who uses the information to make decisions. The customer may also be a Dreamer. The communication motivates and enables those customers to take action.

To fund your Observation project, you will need a research proposal that can be supported by sponsors based on its scientific merit. The Problem Description for an Observation project will succinctly and explicitly outline the current state-of-the-art and identify the gaps in your knowledge. Using the breast cancer example, if you were working in a commercial R&D organization, you might write a Problem Statement for an Observation project as:

The R&D Director needs to understand the genetic causes of breast cancer better because our R&D organization wants to develop a treatment that will reduce the incidence of breast cancer.

THE VISION

DIFFERENT PROBLEMS REQUIRE different solutions. Close-ended problems have a single solution. Most scientific problems, however, are open-ended problems that have many possible solutions; some are likely to be better than others. Problem-solvers use two distinctly different approaches to solve open-ended problems.

Adaptive problem-solving is safe and offers guaranteed results, including good scientific publications. Adaptive problem-solving:
- Makes things better.
- Improves and validates scientific models.
- Improves rules, practices and processes.
- Increases efficiency, eliminates errors and reduces waste.
- Offers incremental improvement to proven principles and practices.

In contrast, radical problem-solving will do things differently. Radical problem-solving reorganizes, restructures or approaches the situation in a new way. Radical problem-solving envisions innovation and creates paradigm shifts[1]. Success leads to inventions, patents, start-up companies and Nobel prizes. But radical solutions have a high risk of failing and failure may lead to bankruptcy and unemployment. Consequently, choose your problems and your problem-solving approach carefully.

Jeff Bezos: If you're not stubborn, you'll give up on experiments too soon. And if you're not flexible, you'll pound your head against the wall and you won't see a different solution to a problem you're trying to solve.

Tool # 5 *The Problem Statement* describes the current situation and states where you are now. Tool # 6 *The Vision Statement* looks into the future and predicts what the world might be like without the problem, in other words, where you want to go. In your role as an Observer, you can turn a solid understanding of the problem into an opportunity that can be captured by implementing the ideal solution. Whether adaptive or radical, your vision should create a sense of excitement, enthusiasm and optimism.

Michael Schrage[2]: Visionary organizations that value innovation should have simple customer vision statements. They need to imagine —and articulate—who and what their customers should become.

Tool # 6 *The Vision Statement* gives you guidance on how to create a vision starting with the Problem Description that details your understanding of the potential causes of the problem. Now, you need to think both strategically and divergently to envision as many ideas for solutions as possible. Tool # 1 *Brainstorming* will help in this process. Use different brainstorming processes to collect ideas. Doing things differently creates discomfort and uncertainty but stimulates people to think differently.

Consider your options and conduct thought experiments[3] to determine the consequences of those potential solutions. Ask yourself "What will the future be like if this problem is solved?" and "Where do we want to go?" In your evaluation of those potential futures use Tool # 2 *Categorical Selections* to consider their relative merits using defined criteria, such as:
- Potential benefits and potential harmful effects.
- Impact and importance.
- Technical and scientific feasibility.
- Consequences of success and failure.
- Risk and opportunity.

Jain, Triandis and Weick[4] observed that the majority of our creative ideas originate from conversations with other scientists. Actively seek conversation and feedback from others. Add their ideas, perspectives and opinions. Revise your vision as you gather more information. Consider advice and feedback from technical experts using Tool # 3.1 *Expert Panel* and Tool # 3.2 *The Delphi Method*. You might also gain different perspectives on the topic using Tool # 3.3 *Deliberative Forum*. Ask specific questions to get feedback. Which option is best? What should I remove? What should I keep? What should I add?

At the end of your envisioning process, you should be able to write a Vision Description document that describes different future scenarios, details the potential solutions, the results of your thought experiments and your reasoning to select the best approach. Your document should list the alternatives that you have considered and the reasons why you have selected this particular option as your vision.

To capture the essence of your vision, write a one sentence Vision Statement. This sentence states what your customer will be able to do that will solve the problem and thereby create a better future. Your Vision Statement envisions only WHAT the future will be like and the associated benefits of a solution. HOW it will be solved comes later. Your Vision Statement communicates a value and an opportunity to potential customers, stakeholders, research team members and decision-makers to gain their support.

If [CUSTOMER] is able to [DO WHAT] then [SOLUTION].

Remember the previous example of breast cancer; the Problem Statement was

Medical doctors need a treatment to reduce the incidence of breast cancer because more than 250,000 women are diagnosed with breast cancer every year.

A potential Vision Statement might be:

Medical doctors will use CRISPR-Cas9 technology to edit mutations in the BRCA1 or BRCA2 genes reducing the incidence of breast cancer from the current 250,000 women diagnosed each year.

Note that the treatment has become more specific and is now CRISPR-Cas9[5] technology. This type of refinement is to be expected in the iterative process of defining a research concept.

Keep your Vision Statement bold and exciting. Avoid drifting into statements on project objectives or identifying needs. Focus on the big picture that captures people's imagination. The Vision Statement is your dream of the future.

Two additional questions are particularly important for you to consider at this time.
- Who will consider your vision to be beneficial?
- Who will consider your vision to be harmful?

Not everyone will view an innovation positively. With the implementation of any change, there are winners and there are losers. Many people simply dislike change, whereas others embrace it. You need to identify these individuals and groups to anticipate either their resistance or their support of your vision. Always consider the emotional and ethical implications. Six Thinking Hats[6] is a useful process to view your vision from several perspectives.

Tool # 3.3 *Deliberative Forum* will identify tensions that may create resistance and support for your vision. Force Field Analysis[7] based on the work of Kurt Lewin[8] can be used to rate the opposing forces or tensions based on their relative strengths and complexity. Based on the feedback from interested stakeholders, you can modify your vision to gain more support.

Tool # 5 *The Problem Statement* and Tool # 6 *The Vision Statement* give a firm conceptual foundation for your research project. These statements will be touchstones for you to reference periodically throughout all subsequent stages of your project. Inspiring Vision Description and Vision Statement documents are exceedingly difficult to write. You require inspiration for good ideas, as well as skilled communication to articulate and promote your vision. These touchstones can be revised and only become written in stone when the sponsors commit funding to the research project. The more detailed analysis written in a Vision Description will be included in legal, business and scientific reviews, and in grant applications. Use your one-sentence Vision Statement as a communication tool to inspire others as you recruit sponsors and research team members.

As an exercise, write a one-sentence Vision Statement for your current research project: *If [CUSTOMER] is able to [DO WHAT] then [SOLUTION]*. You might also ask others on your research team to write their Vision Statement. Do this independently, without cross referencing. Do they have the same customer? Are they trying to solve the same problem? Do they envision the same solution? If you find that their Vision Statements are different from yours, your colleagues are working on a different research project than you are.

LITERATURE REVIEWS

YOUR LITERATURE REVIEW[1] must be more than a descriptive history extracted from previous publications and issued patents, and more than a listing of the observations that others have made. A traditional literature review relies heavily upon the interpretation by the authors of the published article. As such, a simple summary of the literature can be subjective, selective and biased. Instead, your literature review must be a critical, interpretive assessment of current knowledge, methods, statistical analysis and data interpretation related to your problem and vision.

> *Karl Popper[2]: If we are uncritical we shall always find what we want: we shall look for, and find, confirmations, and we shall look away from, and not see, whatever might be dangerous to our pet theories. In this way it is only too easy to obtain what appears to be overwhelming evidence in favor of a theory which, if approached critically, would have been refuted.*

Literature reviews are fallible. Chalmers[3] points out that the observational and logical basis for science is less secure than we might assume. The Dreamer uses abductive logic guided by experience, skill, beliefs, biases and intuition to filter the observations and mold them into a scientific model. Several issues limit one's ability to mold facts logically into a scientific model. Different dreamers see the same event differently because individuals have different perspectives and experience. Different dreamers convert observations into statements of fact differently. What appears to be an important observation to one is irrelevant to another. Filtering data is a skill that requires insight and perception but can be perilous. Judgements about accuracy of observations depend upon what is already known or assumed. In other words, the accepted paradigms of the scientific discipline determine what is considered to be a valid observation and what is considered to be a technical error. Therefore, the observations included in scientific models are fallible.

According to Chalmers[4], even if the data from an experiment are repeatable, the interpretation and significance of the data may change over time for a variety of reasons:

- If your beliefs change, the filters that you use to select data change.

- If an accepted scientific paradigm changes, the context in which you consider the data changes.
- If more accurate or more relevant measurements are made, the historic data become irrelevant.

Sean Covey[5]: Paradigms are like glasses. When you have incomplete paradigms about yourself or life in general, it's like wearing glasses with the wrong prescription. That lens affects how you see everything else.

The quantity of scientific information in many disciplines can be overwhelming. A traditional literature review may require a long time to complete and may be out of date when finished. Searching through the massive amount of scientific information can be exceedingly time consuming. Because a scientist's time is limited, hypotheses can be biased simply because some information was missing from the literature review. Online tools and catalogs of published work greatly facilitate doing a thorough literature review. Reference management software[6] is essential to archive the publications of interest and retrieve electronic copies. Other tools are now becoming available to extract and compile information from the scientific literature in a way that integrates this information to predict relationships. For example, Knowledge Integration Toolkit (KnIT) was a collaboration between Baylor College of Medicine[7] and the IBM Watson[8] team that used computational tools to suggest new relationships and new functions of proteins that modify a tumor suppressor protein[9].

As an alternative to the traditional literature review, many scientists are now conducting meta-analysis[10] to combine results from different studies in an effort to identify and quantify relationships. Meta-analysis can be considered to be a specialized type of literature review that examines data from previous research, ignoring the authors' conclusions. Meta-analysis applies powerful statistical analysis to data merged from many studies. This gives a much larger sample size and is more likely to generate trends and relationships without the same bias as a traditional literature review. Meta-analysis allows a much more quantitative analysis by highlighting correlations among studies without the interpretations that are included in published reports.

The challenge of meta-analysis is the potential for error based on the inclusion of skewed or poor-quality data. The inclusion of a poorly conducted study can jeopardize the entire meta-analysis. Improved database programs have made data entry and extraction much easier and more standardized. Standardized analytical platforms, such as microarray[11] data, have enabled linkages to be made across many experiments, which are compiled into online databases that can be queried from your desktop.

In many disciplines, including medical research, research studies are being assessed and condensed into databases. Systematic reviews[12] by experts that critically evaluate previous studies in a comprehensive way are being published in specialized journals.

The observations and data from a literature review can be molded in several ways into information that is more easily synthesized, understood and incorporated into a scientific model. The organization of data into information that can be understood is arduous. Tools # 7.1 to 7.4 are suggestions that may help with this task. The Is/Is-not Matrix[13] (Tool # 7.1) organizes background data on a relatively straightforward topic. The Four Windows technique[14] (Tool # 7.2) summarizes relatively simple interactions among two independent variables (causes) that have multiple effects. An Ishikawa diagram[15] or fishbone diagram (Tool # 7.3) is useful for visualizing multiple causes of a single effect. A tree diagram[16] is a variation that is suitable to summarize relationships when there are several levels of sub-topics in a hierarchy. In computational biology, a dendrogram is used to represent the relationship among genes or samples. A phylogenetic tree is used to show the inferred evolutionary relationships among various biological species. Whereas a fishbone diagram is useful for single effects, an interrelationship diagram[17] (Tool # 7.4) is useful for visualizing more complex phenomena, in which multiple effects are attributed to multiple causes. A mindmap[18] can be adapted easily to show relationships. The visual nature of the mindmap allows feedback, discussion and easy revision by a team. Creation of a mindmap is well suited to a brainstorming session because it can be easily modified in a visual manner. It is also easily adapted to show all of the other forms of relationships using readily available software[19]

Knowledge about cause-effect relationships is critical to understanding a scientific phenomenon and essential to implement an innovation that solves a problem. Scientific models describe cause-effect relationships among observations and use these relationships to make predictions. You can summarize these relationships, whether proven or hypothetical, using visualization tools that cluster observations into themes.

A cause-effect relationship can be considered proven only if the observations meet three criteria:

1. [CAUSE] is always present when [EFFECT] is observed; there is a correlation between the two variables.
2. The [EFFECT] is observed only when [CAUSE] is present; when [CAUSE] is applied [EFFECT] subsequently occurs.
3. The [CAUSE] is sufficient for [EFFECT] to occur; no other factor is necessary.

These are stringent criteria. In many situations, it is impractical or unethical to conduct experiments to establish these relationships conclusively. Often you must proceed with tentative cause-effect relationships that rely on limited experimental proof. You will also identify situations when separate research groups advocate apparently contradictory relationships. In other situations, relationships are assumed based on correlations (type 1 observations) that are vulnerable to reverse causation errors[20], where the presumed cause is actually the effect. You may also assume relationships based on time-course studies (type 2 observations) that may be spurious. Be sure to flag those relationships as assumptions. Type 3 observations are the most difficult to establish. There may always be something else involved that was not measured or detected in the experiments. On your cause-effect diagrams, you can indicate the confidence that you have in a relationship by using different colors to connect the variables. Alternatively, you can use a letter code, such as the one described in Tool # 7 *The Scientific Model*.

THE SCIENTIFIC MODEL

A SCIENTIFIC MODEL is a fundamental requirement of the scientific method[1]. Without a validated scientific model, it is impossible to conceive a research project. Your summary of the scientific background and technical information on the topic from your literature review is incorporated into a scientific model. Ideally, your scientific model is sufficiently detailed to enable you to solve the problem and implement your vision. A scientific model is analogous to a street map in that it provides an approximation of the relationships among the observations or variables. You need a different level of detail or a different perspective in your street map depending on whether you are walking, cycling, driving or taking the subway to your destination. Likewise, you need different types of scientific models depending upon your problem and your vision. For example, in the life sciences, you require different scientific models depending whether you envision genetic, biochemical, pharmacological or ecological solutions to a problem.

The scientific model may be conceptual, or it may be a mathematical attempt to explain how things work. Conceptual models, such as Bohr's model of the atom, Mendel's concept of a gene or the double helix model for the structure of DNA, simplify complex abstract relationships. On the other hand, mathematical models enable quantitative predictions of complex processes, such as biochemical reactions or crop yields.

LOGICAL ARGUMENTS

Philosophers debate which of several forms of logic should be used to construct scientific models and theories[2]. Ideally, you will use inductive reasoning to extrapolate from your observations that were made on a sample to a generalized conclusion about the larger population. ONLY IF your inductive argument is built on a large number of observations, and ONLY IF those observations have been repeated under a wide variety of conditions without conflict or exception, will you have confidence in your conclusion.

Even so, that confidence may be misguided. Philosophers have noted that it is difficult to use an inductive argument to construct a rigorous scientific model that will withstand criticism. How many observations are sufficient? What constitutes a wide variation in conditions? Were the experiments, treatments and measurements done correctly? The answers to these questions will vary among scientific disciplines and among scientists who have different opinions of what is relevant and important. Chalmers[3] concludes that both observation and inductive reasoning are biased by the beliefs and experience of the scientist, and therefore are fallible. Furthermore, when crafting your scientific model, you will never have all of the data; you will never make all possible observations; and the empirical data that you collect may have inherent variability among subjects and among measurement methods, especially in biology. Therefore, it is a myth that scientists use inductive logic when creating scientific models.

Scientists actually use abductive logic[4] to create scientific models. Abductive reasoning develops the most probable explanation for an incomplete set of observations. We are familiar with abductive reasoning because we use it in our daily lives to make countless decisions. Doctors use abductive reasoning to make a medical diagnosis based on a set of symptoms. A jury uses abductive reasoning to decide whether the prosecution or the defense has the best explanation for the evidence.

Abductive reasoning can make a best guess of an explanation for the observations, even in ignorance of crucial variables. Abductive reasoning produces an inference, not a guaranteed logical conclusion. Many scientific phenomena cannot be explained absolutely or categorically, but only by probabilities. Your success in using abductive reasoning to craft a scientific model depends on your experience, skill, perspective and intuition. Consequently, if you and a colleague review the same data from a set of experiments independently, it is likely that you will craft different scientific models to explain the observations.

Neil Gershenfeld: The most common misunderstanding about science is that scientists seek and find truth. They don't — they make and test models.... Making sense of anything means making models that can predict outcomes and accommodate observations. Truth is a model.[5]

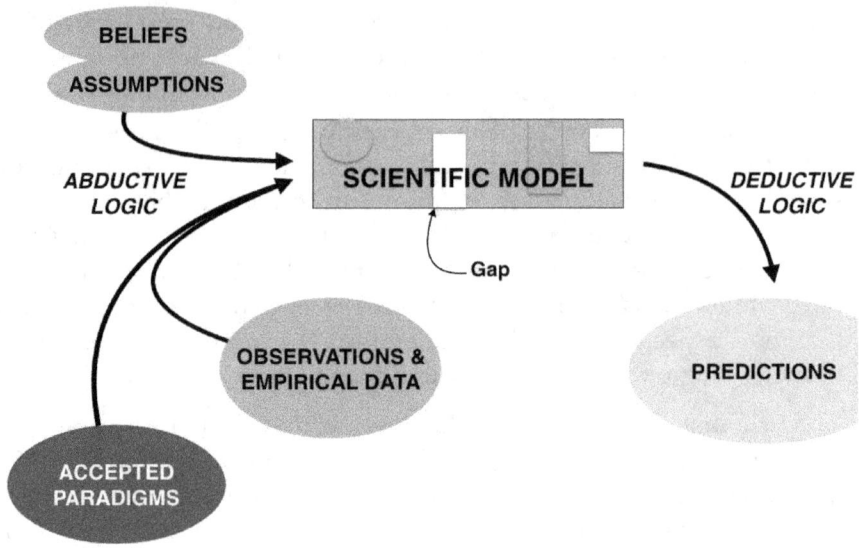

REQUIREMENTS AND CHARACTERISTICS

Popper[6] encouraged Dreamers to make bold hypotheses based on scientific models that can be falsified by subsequent experiments. He considered a hypothesis to be bold if it extends and broadens our knowledge to enable new approaches. Popper contended that knowledge can only be created by the confirmation of a bold hypothesis or by the falsification of a conservative hypothesis. The Dreamer must be creative and imaginative when creating scientific models without making random guesses. Although wild guesses may be bold, they will fail to create knowledge when falsified. Falsification will simply indicate the Dreamer's lack of skill. A classic example is Trofim Lysenko who as director of the Institute of Genetics within the USSR's Academy of Sciences promoted strong anti-Mendelian doctrines that contributed to widespread famine in the USSR and China[7].

Once you have compiled, analyzed and interpreted the information from a literature review, meta-analysis and systematic review and merged it with the commonly accepted theories, paradigms and assumptions using abductive logic, your scientific model can make predictions. However, is your scientific model sufficiently rigorous to make predictions that merit further consideration? Is your scientific model useful in your problem-solving quest? Here are some guidelines to critically assess the quality of your scientific model.

Your scientific model should be useful. The model should propose a path forward by suggesting hypotheses that can be tested or by identifying knowledge gaps that must be filled. If a scientific model contains untested assumptions, the Explorer may choose to test hypotheses based on its predictions. If a scientific model has gaps or provides an incomplete explanation of the previous observations, the Dreamer may choose to refine the scientific model by conducting thought experiments[8] or to become an Observer to collect more data. Only if your scientific model was validated previously by the confirmation of several predictions and only if there are no gaps in knowledge or inconsistencies among observations, will a Builder have sufficient confidence in the model's accuracy, reliability and robustness to use it to develop a prototype of a product.

Your scientific model should emphasize simplicity. The model that explains the most empirical data with the fewest number of assumptions is the better option. The Principle of Parsimony[9], more commonly known as Occam's Razor, states that among competing explanations that predict equally well, the explanation with the fewest number of assumptions is most likely correct. This principle is based on the observation that the accuracy of an explanation increases as its complexity increases, but only to a point, afterwards it declines because greater complexity adds increasing amounts of noise and variability in measurement.

> Karl Popper[10]: *Science may be described as the art of systematic over-simplification — the art of discerning what we may with advantage omit.*

Your scientific model should incorporate current paradigms. The proposition of a new paradigm that rejects a previously held one will create what Kuhn[11] calls a scientific revolution. For example, Darwin's proposal of natural selection rejected the paradigm that species were fixed by divine creation. This irreconcilable inconsistency caused Darwin to delay publication for a decade and as he anticipated, once it was published, it created controversy and debate. When faced with a controversy that involves the interpretation of observations, people's beliefs play an important role in their choice. Therefore, if you are trying to gain the support of sponsors, customers and stakeholders for your research project, recognize the importance of their beliefs and biases in the decisions that they will make. Challenging people's beliefs can be a major source of resistance to your proposal and to the acceptance of your vision.

> Max Planck[12]: *A new scientific truth does not triumph by convincing its opponents and making them see the light, but rather because its opponents eventually die, and a new generation grows up that is familiar with it.*

Your scientific model should explain all of the empirical observations and data. Previous observations that have been published in the scientific literature likely tested relationships among some of the variables in your scientific model and created data, measurements or observations. You can use that previous data to test your scientific model by comparing your model's predictions with these previous empirical observations. Considering the specific experiment's conditions and treatments, what outcome would your scientific model predict based on deductive logic? The challenges associated with making conclusions from the predictions of a scientific model are discussed in the Duhem–Quine thesis[13] which argues that predictions from a scientific model require several assumptions about relationships.

If the previous empirical observations are in agreement with your prediction, then you have validated your scientific model under those specific conditions. However, you have not proven it under all possible conditions. Furthermore, Chalmers[14] and other philosophers argue that the truth of a prediction cannot confirm the truth of the premises. On the other hand, if a previous empirical observation is inconsistent with the prediction from your scientific model, you may conclude that the lack of consistency is because:

One (or more) of the premises used in your logical argument is wrong. Therefore at least part of your scientific model is wrong. But you cannot identify which premise is wrong.

OR

The experiment was done incorrectly. Errors were made in the treatments or in the measurements. In which case, the premises are true, but the experiment was flawed. Therefore, the scientific model may be correct.

Most scientists chose the second option as their default conclusion, criticizing the technical skill of others and discounting the erroneous observations.

Your scientific model should be dynamic. Models start by being overly simplistic and become more refined as more observations and new interpretations are added. Your first scientific model will be incomplete. Black boxes may represent unknown concepts. Gaps in your knowledge may limit your understanding. Many new terms, such as atom, electron, quark and gene, were included in scientific models to explain empirical data, even though no one had ever seen an atom, electron, quark or gene. Scientists had simply inferred their existence in a scientific model to explain observations. You may need to craft similar concepts to cover the gaps in your scientific model.

Your scientific model should withstand criticism. Use Tool # 3 *Review Panel* to obtain expert and peer review of your scientific model. You may also seek external peer review of your analysis, ideas and concepts by publishing in a peer reviewed journal. This publication will provide even more feedback from reviewers before publication and from peers after publication. This publication will establish a benchmark to enable you to compare your ideas with others. However, a word of caution, publication of your ideas may prevent claims to intellectual property because your publication will be cited as prior art in any subsequent patent application. For this reason, scientists in commercial R&D organizations publish only after submission of patent applications.

Tool # 7 *The Scientific Model* provides a guideline on how to summarize the current state of knowledge on your specific scientific phenomenon and construct a scientific model. A high degree of craft, skill and art are prerequisites. The scientific model that you create is based on Tool # 5 *The Problem Statement* and Tool # 6 *The Vision Statement*. Your scientific model should provide a road map on how to achieve your vision and solve your problem. Now that you have a better understanding of the scientific literature and have drafted a working scientific model, you may need to revise both the problem and vision statements before proceeding to draft research objectives.

GAPS IN KNOWLEDGE

Gaps in knowledge may simply be an inconvenience or they may be absolutely critical to the implementation of your vision. The Dreamer asks: What is the significance of these gaps to the scientific model? What is the significance of these gaps to the vision?

If the gap in your knowledge is important, write a Gap Statement.

You will need to fill in these knowledge gaps by conducting an Observation project to collect more data.

If we knew [WHAT], [CUSTOMER] would be able to [DO WHAT].

Remember the previous example of a Vision Statement:

Medical doctors will use CRISPR-Cas9 technology to edit mutations in the BRCA1 or BRCA2 genes reducing the incidence of breast cancer from the current 250,000 women diagnosed each year.

Your literature review will have identified several potential sites of mutations in two genes, BRCA1 and BRCA2, that have been linked to breast cancer. The gaps in your knowledge can be summarized by several Gap Statements:

1. *If we knew that editing the BRCA1 or BRCA2 mutations would prevent the formation of tumors, then medical doctors would be able to prevent breast cancer in those patients who have a genetic predisposition.*

2. *If we knew how to target CRISPR-Cas9 to specific precancerous cells, our R&D organization would be able to develop gene editing technology into a treatment.*

3. *If we knew how whether the CRISPR-Cas9 treatment of BRCA1 or BRCA2 mutations had any unintended effects, the FDA would be able to assess the safety of the gene editing technology.*

Writing a succinct Gap Statement is an iterative process (Tool # 7.5 *The Gap Statement*). Avoid the temptation to do this alone. Instead, recruit a team to revise and comment on your ideas, or better yet, use Tool # 1 *Brainstorming*. Some gaps may be filled with existing knowledge from others. Confirm your analysis using Tool # 3.1 *Expert Panel* or Tool # 3.2 *The Delphi Selection Method*. Preliminary Gap Statements like those above provide a starting point for information sharing and discussion. Some gaps may remain unknown. You will need to decide whether to fill those gaps or proceed based on an assumption. Your scientific model and the identification of your gaps in knowledge will guide you in conceiving the appropriate Observation, Modeling, Discovery and Development projects.

PREDICTIONS

Your scientific model predicts relationships among the component variables of the scientific phenomenon that you are studying. Some of those relationships are established by empirical data and others are assumed with varying degrees of certainty. The prediction of a relationship among two or more variables in a scientific phenomenon is a logical argument[15]. This philosophical argument asserts or denies that the members of one group of variables are included in another. Your scientific model allows you to predict something about their relationship. You use deductive logic to go from a generalization in the scientific model to the outcome in a specific circumstance, which is a hypothesis. A hypothesis allows you to predict the results of an experiment and therefore can be tested by empirical data. From Tool # 7.6 *The Prediction Statement*, a simple one-sentence statement will provide focus, clarity and motivation for you and your research team as you develop the concept for a research project.

Because [A] causes [B] then [X] will occur.

For more complex problems you might write:

Because [A] causes [B], and because [C] causes [D], and because..., then [X] will occur.

Using the previous example of breast cancer, a one-sentence Prediction Statement might be:

Because the BRCA1 gene functions in DNA replication and repair, mutations of the gene will cause a loss of the DNA repair function leading to the formation of tumors.

This Prediction Statement is based on assumptions and identifies knowledge gaps:

Assumption 1: loss of function in DNA repair leads to the formation of tumors.

Assumption 2: the BRCA1 gene encodes a protein that is essential to repair DNA.

Knowledge gap: the specific site mutations in BRCA1 that cause loss of the DNA repair function.

This Prediction Statement therefore frames the objectives for several potential research projects. It is your decision whether to accept the assumptions or not, and whether to fill the knowledge gap or not. The assumptions may be rewritten as hypotheses to be tested in a Discovery project. The knowledge gap may be addressed in an Observation project. You, or preferably your research team, must decide which assumptions to accept, which hypotheses to test and which knowledge gaps to fill. However, you should do this explicitly with full knowledge and disclosure of the assumptions that you have accepted and those that you have chosen to test.

YOUR SCIENTIFIC MODEL

As an exercise, write or draw the scientific model for your current research project. Despite the fact that they are difficult to construct, I suggest that you use a diagrammatic, iconic or pictorial representation of cause-effect relationships to communicate complex topics to customers, stakeholders, research team members and decision-makers. A graph communicates a relationship more clearly to most people than an equation. A story board communicates summary information better than 20 pages of text. For example, to show the mode of action of a certain chemical, six story boards may be organized: molecular, metabolic, cellular, tissue, organism and ecological effects of the chemical can be drawn on separate boards. A flow diagram depicts a sequential process that can identify gaps, missing knowledge, alternative paths or unnecessary steps. A biochemical pathway is an example of a scientific model comprising a series of cause-effect relationships presented as a flow diagram.

Write the Gap Statements that you are using in your research project. *If we knew [WHAT], [CUSTOMER] would be able to [DO WHAT].* Write the Prediction Statements that you are using in your research project. These are not testable hypothesis statements, although they may develop into hypotheses. What are the logical predictions from your scientific model? Practice using deductive logical arguments. **Because [A] causes [B] then [X] will occur.**

MODELING PROJECTS

IF YOU FOUND the previous statements difficult to articulate, then perhaps you need to conduct a Modelling project. Your literature review of the published scientific information on a phenomenon may, in some cases, lack the insight that you need to make predictions. Modeling projects address problems that lack a scientific understanding about interactions and cause-effect relationships (Category B problem). As a Dreamer, you want to predict future events. A Dreamer uses creative thinking, abductive logic, analogical reasoning, experience, skill, and technical knowledge to improve a scientific model[1]. A Dreamer analyzes data, facts and information. A Dreamer digs deeply, understands the relationships among the components of a phenomenon and uses divergent thinking to consider all options and possibilities. The data published by Observers may be inconsistent; some of the data may be unreliable; some of the data may be irrelevant; some of the data may be misinterpreted. Meta-analysis of published data may provide a better insight. Another option is to conduct thought experiments[2] to find potential relationships. The Dreamer filters and molds knowledge using abductive logic into a scientific model that explains the empirical observations. The model provides the simplest and most likely explanation for the observations. The model predicts future events, creates options and suggests hypothetical solutions to the original problem.

> *Moti Ben-Ari[3]: A scientific theory is a concise and coherent set of concepts laws and claims; precisely and accurately explains and predicts natural phenomena; should include a mechanism preferably mathematical that explains how its concepts claims and laws arise from lower level theories.*

The Dreamer is a creative thinker. Divergent thinking[4] processes create ideas, options and alternatives by reorganizing information in new ways to improve scientific models. Convergent thinking[5] processes sort through those options to find a unifying explanation. The Dreamer has extensive experience and uses analogical reasoning to construct scientific models by borrowing ideas and relationships from one scientific phenomenon and applying them to a similar phenomenon. The Dreamer's skill and knowledge guide the identification of phenomena which share common characteristics and relationships.

Charles Darwin and Gregor Mendel are historic examples of Observers who became Dreamers. Dreamers, who took their observations or empirical data, respectively, from Observation projects and molded fundamental scientific models of natural selection and heredity. At the time, both projects would have been considered to be basic research, but both enabled enormous problem-solving applications. Category B problems are not solely basic research problems. For example, Robert Fraley conceived and advocated a scientific model for herbicide resistance in crops that enabled Monsanto to develop the first commercial glyphosate resistant soybeans and corn crops, which, in turn, has revolutionized crop production in the United States[6].

A preferred option is to create rival scientific models. These rivals explain the observations with equal rigor but propose different cause-effect relationships. The advantage of this approach is that rival scientific models propose rival hypotheses that can be experimentally tested. Kuhn[7] proposed that it is impossible to falsify one scientific model without validating its rival. A new model would be accepted by the scientific community only if it explains more of the observations and only if it provides more potential solutions to problems and puzzles than the original. If the predictions of a scientific model are refuted without an alternative model, then the most likely conclusion is that the experiment was flawed. In contrast, if the predictions of one scientific model are refuted and the predictions of another confirmed, knowledge about a scientific phenomenon has been created leading to a better understanding and a better scientific model.

Karl Popper[8]: Whenever a theory appears to you as the only possible one, take this as a sign that you have neither understood the theory nor the problem which it was intended to solve.

A Modeling project is often a continuation of an Observation project if new measurements or new relationships challenge a scientific model. The improved or rival scientific model is communicated in a scientific publication and generates hypotheses (predictions) that can be tested by

The Research Proposal

experiments as part of a subsequent Discovery project.

Modeling projects are supported by sponsors based on scientific merit. The customers include a diverse group of Explorers who test the predictions of scientific models. These customers also include teachers and students who seek understanding and Builders who seek innovation.

Dreamers think creatively. If you aspire to be a successful Dreamer, you must create lots of ideas, options and alternatives by reorganizing information in new ways. Perhaps you need a new scientific paradigm. Perhaps you need a radical innovation. Do not believe that you must do this alone; seek help from others. Tool # 1 *Brainstorming* will help you in this task. Be assured, after a lot of hard work, you will find a unifying explanation and create a better scientific model. To fund your Modeling project, you will need a research proposal that can be supported by sponsors based on its scientific merit.

KEY POINTS

A RESEARCH PROJECT acquires the knowledge needed to solve a problem.

Problems for research projects come in many forms:

A. Problems that are poorly described require more observation and information.

B. Problems that lack understanding require better scientific models.

C. Problems that have several hypothetical solutions require testing.

D. Problems that need validation of a solution require implementation of an innovation.

Tool # 5 *The Problem Statement* describes a process to define why a problem is important to solve. The Problem Description provides technical details that will be included in business proposals and in grant applications. The Problem Statement is a one-sentence communication tool: **[WHO] needs [WHAT] because [WHY]**.

An Observation project seeks to characterize a problem better, to view it from a variety of perspectives and to look at all elements that may contribute to it.

Different problems require different types of solutions. Adaptive solutions make things better by improving scientific models, rules, practices and processes. Radical solutions reorganize, restructure or approach the problem in a new way, shifting scientific paradigms, creating new scientific models and driving innovation.

Tool # 6 *The Vision Statement* uses strategic and divergent thinking to look into the future and to envision opportunities. The Vision Description is used in grant applications and business proposals. The Vision Statement is a one-sentence communication tool: **If [CUSTOMER] is able to [DO WHAT] then [SOLUTION]**.

To test the feasibility of your vision, ask: Who will consider your vision to be beneficial? Who will consider your vision to be harmful?

A scientific model is a fundamental requirement of the scientific method on which research concepts, hypotheses, experiments and conclusions are based.

Tool # 7 *The Scientific Model* summarizes current knowledge, expands our understanding of a scientific phenomenon, enables opportunities to be envisioned and formulates hypotheses to be tested. This involves a high degree of craft, skill and art.

A scientific model should:
- Envision opportunities.
- Communicate a concept.
- Emphasize simplicity.
- Incorporate current paradigms.
- Explain all of the empirical observations and data.
- Be dynamic.
- Be useful.
- Withstand criticism.

Tool # 7.5 *Gap Statement* identifies incomplete explanations of the data and assumptions that have not been validated: **If we knew [WHAT], [CUSTOMER] would be able to [DO WHAT].**

Tool # 7.6 *Prediction Statement* uses deductive logic to propose relationship among established observations (premises): **Because [A] causes [B], then [X] will occur**.

A Modelling project may be required to refine or improve a scientific model before it can make predictions. An Observation project may be required to fill the knowledge gaps in the scientific model before it can be used to make predictions.

PART FIVE

DRAFTING A RESEARCH PROPOSAL

THE OBJECTIVE

YOUR SCIENTIFIC MODEL should predict one or more potential solutions to the original problem that will achieve your vision. The solution may be challenging and may require your ingenuity, your hard work and all of your skill to implement, but it is theoretically possible because it is consistent with current knowledge and previous experience. Tool # 7 *The Scientific Model* allowed you to write both a Gap Statement and a Prediction Statement. The next step is to add a specific attainable objective for one or more research projects. If at all possible, do this as a team exercise. The team provides diversity of technical knowledge and skill, as well as different perspectives and beliefs.

> *Lewis Carroll (reworded from Alice in Wonderland)*[1]: *If you don't know where you're going, any road will get you there.*

All research projects have an objective to learn something. Their specific objectives differ depending upon the type of problem, your vision and the status of your scientific model. Achieving the objective is the endpoint of a research project and defines success. A clearly defined objective empowers a research team to make decisions and provides focus. An objective helps the stakeholders, research team, customers and sponsors to align their expectations.

To determine which knowledge gaps that you should address in your research project, begin by listing your options. Each of the options becomes a potential research project. Tool # 1 *Brainstorming* is a good initial approach to gather these options, followed by a convergent thinking selection process (Tool # 2 *Categorical Selections*) including a technical evaluation (Tool # 3 *Review Panel*) to create a short list. Make a final selection as a team using Tool # 4 *Group Voting*. Tool # 8 *The Objective Statement* produces an Outcome Description that indicates what knowledge must be created, why this knowledge is needed, what will change afterwards, as well as any other relevant information. The statement details the accomplishments that will validate the predictions of your scientific model, for example, a cause-effect relationship that will fill a knowledge gap. Note that these are accomplishments, not activities. The Outcome Description does not state how you will do something, but only what you will accomplish. Your Outcome Description can be used in business proposals and grant applications.

Now condense and squeeze the Outcome Description into a one-sentence Objective Statement. You may find that you have several objectives. In which case, you may consider organizing the work into several research projects as part of a research program. For example, here are some generic objective statements for different types of research projects:

1. An Observation project learns something that is revealing and provides more understanding about a problem.
   To make more precise measurements, i.e. more quantitative and less qualitative.
   To make observations in different situations or with different treatments.
   To make different measurements using new technologies or methods.
2. A Modelling project learns about relationships, develops a better scientific model and proposes new solutions.
   To resolve previous anomalies, ambiguities or inconsistencies.
   To develop a universal law or a physical constant that unifies observations.
   To make predictions of relationships among components or attributes.
3. A Discovery project learns whether hypotheses, predictions, relationships and assumptions are true.
   To test hypotheses that select among rival scientific models.
   To select the best formulation or method.
   To validate performance of a product or technology in new conditions.
4. A Development project learns whether your knowledge, discovery or invention will implement a solution to a problem.
   To produce a prototype that demonstrates the utility of an invention or discovery.
   To produce factual information that creates value (e.g. marketing or efficacy data).

Here is a classic example of an objective that is often cited in the project management manuals as a model:

> *John F. Kennedy (1961): I believe that this nation should commit itself to achieving the goal, before this decade is out, of landing a man on the moon and returning him safely to earth.*

Kennedy's objective sparked many research projects that created inventions, new technology and major scientific advances including transistors, micro-electronics, medical monitoring devices and telecommunications devices. These innovations put a man on the moon but also changed the way we live, the way we do business and especially the way we communicate. These changes may have occurred without Kennedy's statement, but more slowly and more haphazardly. He empowered people to act and to make the necessary decisions to achieve his stated objective. Your Objective Statement should do the same.

The Objective Statement of a traditional construction-like project states WHO will do WHAT by WHEN. The project management manuals state that a project's objective should be:
- Succinct.
- Simple and easily understood.
- Time-specific.
- Challenging but possible.
- Defined by success criteria that can be measured transparently.

Tool # 8 *The Objective Statement* will guide you through a process to write an objective for a research project in one sentence. Be succinct and precise. Brevity forces clarity.

Our team will learn [WHAT].

There are some optional components that may be included in your Objective Statement to improve clarity. You can replace "our team" and specify exactly WHO will conduct the research project, but at an early stage of conception, this may be premature. Your Objective Statement will likely be used to recruit team members. Therefore, replace "our team" in your Objective Statement with a specific entity, group or individual only when responsibility has been established.

This statement differs from the classic Objective Statement because it simply states what you will learn. You can and should specify a deadline to do something. It is more problematic to specify a deadline to learn something. Admittedly research grants have a fixed duration; student theses have deadlines; research projects end. However, research projects in commercial and government organizations have a fixed scope and quality, in which cases time cannot be fixed according to the triple constraint principle. Therefore, include WHEN in your Objective Statement only if there is a fixed duration to the project. If you do so, avoid fuzzy wording in your objective that hedges against the deadline.

Avoid the temptation to overextend yourself and exaggerate what you may learn in your Objective Statement. Your enthusiasm and long-term vision needs to be moderated or you will create false expectations in your customers. This will sow the seed for disappointment and frustration. So set a challenging goal but be realistic because you will be held accountable.

> *Dalai Lama: No matter what our motivation may be, if we are not realistic we will not fulfill our goal[2].*

As an example, remember the Problem and Vision Statements concerning breast cancer. The first Gap Statement was:

If we knew that editing the BRCA1 or BRCA2 mutations would prevent the formation of tumors, then medical doctors would be able to prevent breast cancer in patients who have a genetic predisposition.

This knowledge gap concerns the postulated cause-effect relationship between mutations of the BRCA1 or BRCA2 genes and the formation of tumors leading to breast cancer. Therefore, an Objective Statement might be:

The BRCA team will learn whether a mutation in either the BRCA1 or BRCA2 genes is a potential cause of tumor formation leading to breast cancer.

Note that the Objective Statement simply states the knowledge to be learned. The strategic and tactical planning of HOW this will be learned is best done by the respective project team that must actually do the work. Therefore, avoid stating HOW anything will done in the Objective Statement.

Because the Objective Statement defines the knowledge deliverable that sponsors and customers expect to receive, the Objective Statement for a research project becomes a formal requirement, once the research project is funded. In other words, it becomes a contract! Until then, your Objective Statement is malleable and can be iteratively modified as you develop the strategic and tactical plans to achieve it.

Once funded, the strategy and tactics to achieve the objective may change, but the objective itself remains. If it becomes apparent that an objective cannot be achieved, then the research project, experiment or task should be explicitly terminated, and a new one started. If the project slowly morphs into something with different objectives, this revision creates confusion and consequently dissatisfaction among sponsors, customers and stakeholders. You will also fail to learn how to improve for your next research project.

Individual experiments and tasks within a research project can also benefit by using an Objective Statement. The difference is one of scope, not process, and a modification of Tool # 8 *The Objective Statement* will work well. In those cases, the traditional project management style works well and a time limit should be included; so the wording of the Objective Statement should be:

[WHO] will accomplish [WHAT] by [WHEN].

Be careful that you specify "accomplish" instead of "do". You should focus on success in achieving a milestone, not on doing something. For example, it is better to set an objective to achieve something because you conducted an experiment, than to simply have an objective to conduct the experiment. You will always be successful in conducting the experiment, even if you do not achieve what you had hoped. Setting the objectives with this wording allows you to change your tactics and experimental design, if an opportunity arises to achieve the objective more efficiently. This flexibility helps you subsequently to manage a research project to success, as opposed to manage an experiment to completion.

As an exercise, write an Objective Statement for your current research project: *[MY TEAM] will learn [WHAT]*. Additionally, write an Objective Statement for an experiment within your research project: *[WHO] will accomplish [WHAT] by [WHEN]*. You might also ask others on your research team to write their Objective Statement. Do this independently, without cross referencing. Do they have the same objective? Are they trying to learn the same thing as you are? If you find that their Objective Statements are different from yours, your colleagues are working on a different research project than you are. They have different motivations and different expectations. They will make different decisions. This will create conflict that may become personal.

THE RESEARCH CONCEPT

THE CRAFTING OF a research proposal is an iterative process for the research team. Tool # 9 *The Research Concept* provides a summary of the information compiled so far. Your Research Concept is a merged document containing:

- ✓ The Problem Description for Tool # 5.
- ✓ The Vision Description from Tool # 6.
- ✓ The Scientific Model from Tool # 7.
- ✓ One or more Gap Statements from Tool # 7.5.
- ✓ One or more Prediction Statements from Tool # 7.6.
- ✓ One or more Objective Statements from Tool # 8.

When completed, the Research Concept will be part of a portfolio of concepts. This is a good time to let your ideas for a research project incubate, which is a critical step in Sawyer's Zig and Zag process[1] for creative problem-solving. The iterations continue as you review and further improve your concept.

The Research Concept document will help you to select the most appropriate projects from this pool to submit to a sponsor to seize a funding opportunity quickly. Whether to proceed with a proposal or not is a major decision. The professor must decide whether to submit a grant application to a sponsor. The Research Director in a commercial or government R&D organization must decide whether to allocate scarce resources to plan and conduct a full project. The investor must decide if there is sufficient potential for a return on investment. To enable these decisions, the Research Concept should address the following questions that sponsors and decision-makers in the R&D organization will have:

- Why does this problem merit attention?
- Will this vision provide an acceptable solution to the problem?
- What do we need to learn that will create this vision?
- What do we need to accomplish to create this knowledge?

The Research Concept' accomplishment may be building a prototype, testing a hypothesis or making new observations. List these as explicitly as possible. Then review these accomplishments to ensure that they will actually achieve the objective. As in Tool # 8 *The Objective Statement*, avoid stating what you will do or what experiments you will conduct. These tactical details will be added later. But at the same time, it is important that the objective be realistic; so it is probable that you will have some conception of the type of experiments needed, just not the details. I emphasize again that the concept for a research project is not to do something, but to accomplish something. Many scientists fail to make this distinction in their research proposals.

At this stage, you will guess the project's duration, how much it will cost and what quality of information is "good-enough" to achieve your objective. This guess provides a preliminary estimate of the scope of your research project. You should also consider what your research project will NOT do. Otherwise, your research project may become a never-ending quest. You should consider what work can be done within your R&D organization with your existing infrastructure and what might be done by service providers or external collaborators. Finally, you should consider what knowledge is absolutely critical to achieve the objective and what knowledge is nice-to-have. This latter consideration is useful for budget negotiations with a sponsor or a contractor. Note at this stage that your planning is only conceptual to establish a framework for the research project. You will add technical detail later.

I recommend that you have at least five Research Concepts in your portfolio, ready to propose when funding opportunities arise. As these potential projects incubate in your portfolio, there is an opportunity to prioritize and to improve the concepts to better meet the expectations of sponsors and customers. The following are some supplemental subjective analysis tools to review and to compare alternative concepts from different perspectives, leading to an iterative improvement of your research proposal.

SWOT ANALYSIS

SWOT Analysis[2] is commonly used in commercial organizations to assess projects and opportunities. This method creates a subjective list of the Strengths and Weaknesses that are internal to the R&D organization, and the Opportunities and Threats that are external. Strengths of the R&D organization that support a research project may include skill of research staff, facilities and equipment. Weaknesses in the R&D organization constrain the research project. Opportunities in the external environment will add value. Threats from the external environment, such as the actions of competitors, may impede success.

SWOT analysis will not only analyze but may improve a concept. By recognizing that a weakness exists, steps can be taken to bolster that weakness. The potential success of a project may be improved by matching strengths and opportunities to capitalize on helpful advantages. A threat may be an opportunity depending on one's perspective. Some view the glass as half full, others view it as half empty. Try to convert weaknesses or threats into strengths or opportunities. For example, some might view your lack of intellectual property as a threat because competitors might block development of your product. Alternatively, you might view the lack of intellectual property as an opportunity to file your own intellectual property quickly and capture a new market opportunity. Lisa Furgison[3] suggests to conduct TWOS following a SWOT analysis to develop strategies that will capture helpful aspects of a concept and minimize harmful aspects. In Tool # 9.1 *SWOT Analysis*, I have included some questions that may help you utilize the strengths and opportunities in your concept for even greater benefit.

FORCE FIELD ANALYSIS

A version of Force Field Analysis (previously used in Tool # 3.3 *Deliberative Forum* and described in Tool # 9.2 *Force Field Analysis*) can also be used to check the feasibility of a potential research project by recognizing the factors that are likely to help and those likely to hinder approval of the research project by the decision-makers (e.g. potential sponsors, customers, administrators or business leaders). The tool looks at forces opposing and supporting your project and rates these based on their relative strengths and complexity. Different processes can be used to gather the ratings[4]. Decision-makers may be influenced by several factors, including:

- Their understanding of the concept. Clear communication of the problem, vision and objectives is required.
- Scientific disagreement about the advantages and disadvantages of acquiring the knowledge. Often this is based on disagreements about paradigms, models, assumptions and hypotheses.
- Confidence that the research team will deliver the expected knowledge. This may be based on an assessment of past accomplishments, e.g. academic publication record or outcome of previous projects.
- Their self-interest based on their concern about how the research project or its innovation will change or impact their interests.
- Their tolerance to change, job security and the degree of

organizational stability.

- Ethical or moral issues, especially concerning human subjects or environmental impact.
- Legal or regulatory compliance issues.
- Financial, budget or return on investment issues.

Earlier, you possibly conducted a force field evaluation of your vision to determine possible support and resistance to an idea. Tool # 6 *The Vision Statement* advises you to consider the factors likely to help vs hinder support for your vision. This more extensive force field analysis also includes the scientific aspects of the research concept. How much support will there be for the scientific model and its predictions? Who will support and who will resist? If you are proposing a paradigm shift or a radical innovation, you might anticipate resistance from the scientific experts. Tool # 3.3 *Deliberative Forum* will provide you with some understanding of how others perceive your concept.

Another use of Force Field Analysis is to view a research project from the perspective of the stakeholders who will be impacted by the project or by the innovation created by the project. Is this stakeholder likely to support or oppose this project? Why? Then plan ways to minimize opposition and bolster support from each stakeholder. Another variation of this tool can also be used to look at various future scenarios that may be created by an innovation or change. Some people will support your innovation, whereas others will oppose it.

WIN CONDITIONS

Another approach to review your concept is what DeCarlo[5] calls "win" conditions. He distinguishes between project and personal win conditions. The seven project win conditions are schedule, budget, scope, quality, return on investment, stakeholder satisfaction and team satisfaction. Personal win conditions are those outcomes that are desired by specific individuals, including stakeholders, decision-makers and research team members. At this stage of the drafting the research proposal, it is important to consider the personal win conditions of the key decision-makers that were identified in Tool # 9.2 *Force Field Analysis*.

The schedule win condition means completing the project on time. Research projects that are funded by a grant have a fixed duration and can almost never be extended. However, grants can be renewed after a review, but that review requires a significant accomplishment from the previous grant. Delays are rarely viewed favorably. In commercial organizations, schedule is equally important because the research is often tied to product development which depends on market conditions. Usually the first to market with a new product has a commercial advantage and intellectual property rights. Schedule is also important to an interdependent research team that has handoffs from one task to another, such as a sample preparation handoff to an analysis handoff to a statistical analysis handoff. Schedule may also include meeting deadlines and milestone dates, such as scientific meetings, performance reviews or budget meetings.

The budget win condition means completing the project within budget at the estimated cost. Research projects funded by grants cannot have cost overruns. Research projects in government and commercial R&D organizations can go over budget but this initiates scrutiny that includes lots of potentially difficult questions.

The scope win condition means completing the tasks and acquiring the knowledge that was required or anticipated. Scope is determined by the triple constraints: the time that is allowed, the budget that is approved and the quality of the knowledge required. Two or three of these constraints are commonly fixed in different types of R&D organizations. If any one changes during a project, scope is affected.

The quality win condition means acquiring rigorous knowledge that will withstand critical review leading to publication in a high quality, peer-reviewed journal or award of a patent. Quality also means delivering a prototype based on acquired knowledge that meets specific performance criteria.

The return on investment win condition means that the knowledge acquired in the research project has sufficient value to justify the expense. This is based on the financial expectations of the sponsor (investor) on the economic return from their financial support. Tool # 10 *Financial Analysis* provides some guidance using either return on investment (ROI) or net present value (NPV).

The stakeholder satisfaction win condition means that all of the people involved or impacted by the research project are satisfied. The expectations of the customers who will use the knowledge acquired have been met. The expectations of the stakeholders who manage the R&D organization's resources and the administrators and decision-makers in the R&D organization are satisfied. This often means achieving certain administrative outcomes, such as staying within budget, complying with regulatory and legal requirements and meeting ethical standards. Legal, regulatory, ethical, safety and infrastructure constraints restrict the scope of some research projects. In addition, individual stakeholders will have personal win conditions that must be satisfied as the research project and experiments are conducted.

The team satisfaction win condition means providing benefit to the research team from their participation, for example gaining experience and skill demonstrated in the form of scientific publications that will promote their careers. Satisfaction also means having a challenging and stimulating work environment. Individual team members will have personal win conditions. Graduate students expect to receive a degree that leads to employment. Postdoctoral students expect to receive recognition in the form of high quality publications that establishes their reputation. Research associates expect professional skill development and financial benefits. Professors expect to receive tenure and grant renewals. All expect a challenging, stimulating and safe work environment.

Tool # 9.3 *Win Conditions* guides you through a process to identify the most important win conditions that must be addressed in the Research Concept because they will impact the support of the research team, the sponsors and the customers for the potential research project.

RESEARCH PROJECT SKINNY

The Research Concept will likely be quite large, suitable for inclusion in grant applications and business proposals. Consequently, it will require a lot of dedication to read. Most stakeholders, senior executives, business leaders and potential recruits to the research team lack the time required to read and understand the Research Concept, at least initially. You must first capture their attention and their imagination. A useful way to communicate a Research Concept and capture the interest of these people is the Research Project Skinny. This term was originally coined by Doug DeCarlo[6] as part of eXtreme Project Management. I have modified his wording for a research project. I recommend its use because its brevity forces a research team to state exactly what the research project must do to succeed. Brevity strips away the scientific complexity and the technical jargon that sometimes obscures our intentions and obscures understanding by a layperson. The conversation within the team drafting the skinny creates commitment. The skinny becomes a touchstone to which the research team members can refer when making decisions.

The Research Project Skinny states: *[WHO] will learn [WHAT] for [WHOM]. This research project will be considered finished when [WHAT] has been accomplished. This knowledge will [DO WHAT].*

The first sentence identifies the research team, the knowledge to be delivered and the customer who will receive or use the knowledge. Note that this sentence contains the research objective from Tool # 8 *The Objective Statement* and has been expanded to include the customer who will use the knowledge and to identify the team or individuals who will conduct the experiments. The second sentence states explicitly what will be achieved and lists accomplishments. It is not a list of activities. The third sentence indicates how this knowledge will contribute to the solution of a problem.

If the Research Project Skinny is imposed by a dominating Principal Investigator, the concept may be accepted but individuals will not be motivated. Crafting a Research Project Skinny as a group has several advantages beyond its use as a communication tool. First, the team understands the essential components of the project work:

- WHO will do this work?
- WHAT must be learned?
- WHO is the customer that will use the knowledge?
- WHAT must be accomplished for the customer to consider the research project successful?
- WHAT will this knowledge do? or HOW will the customer use this knowledge?

Discussion builds commitment and motivation within the research team. Interdependencies become more obvious and accepted. Awareness enables delegation of decision-making. Better decisions will ultimately be made during the project as experiments are conducted.

Secondly, different and often new perspectives are presented that may improve the Research Concept. One of those improvements will come when multiple answers are given for the above questions. If different team members have overlapping responsibilities, if they are seeking different types of knowledge for different customers, if the end point of the project is not well-defined, there will eventually be conflict. It is best to resolve that conflict early, before the project even starts. For example, a research project may have multiple customers - a businessman who wants to use the knowledge to promote commercial products, a lawyer who wants to file patent applications, the graduate student who wants to publish scientific papers and submit a graduate thesis, the regulator who wants to establish environmental protection guidelines. Which customer is the most important? Who is the primary customer? Can the deliverables be made in a sequential manner that allows all customers to achieve their expectations or must a choice be made? Should there be multiple projects with different customers? These are important issues to clarify before grant applications are made and before potential customers become dissatisfied. Tool # 11 *Customer Assessment* provides some guidelines.

Finally, the Research Project Skinny will serve as a touchstone during the subsequent planning and execution of the project. This will enable better decision-making and more open communication because the team shares a common agreed goal. When a decision needs to be made, the team members will remember the discussion and make a more informed decision.

The Research Project Skinny for the previous example of a breast cancer treatment might read:

The BRCA team will learn whether a mutation in BRCA1 or BRCA2 gene is the cause of tumor formation leading to breast cancer for the Research Director of our R&D organization.

This research project will be considered finished when we have learned:
- whether a mutation in BRCA1 or BRCA2 gene is present in all cases of breast cancer.
- whether breast cancer tumors occur only in patients with a mutation in BRCA1 or BRCA2 gene.
- whether a mutation in BRCA1 or BRCA2 gene is sufficient to cause formation of a breast cancer tumor.

This knowledge will allow our R&D organization to decide whether to proceed with development of a gene-editing product.

Note that the Research Concept still does not detail HOW the knowledge will be learned, nor does it list the activities or experiments. Instead, it states the knowledge to be learned and the accomplishments that indicate success in acquiring that knowledge.

UNDERSTANDING SPONSORS

THE SPONSORS WHO fund scientific research vary tremendously in their motivations. In some cases, the motivation is philanthropic to benefit society, but in most cases it is for financial or commercial gain. Your sponsors may also be your customers who have a contract with your R&D organization to deliver specific knowledge, but for this discussion I will consider them separately.

Globally, taxpayers, industry, entrepreneurs and dedicated associations invested $1.7 trillion in 2015 for research and development projects[1]. That investment is increasing every year. The combined investments of only 10 countries account for about 80% of 2015's total R&D budget. According to data compiled by the National Science Board, R&D organizations in the USA invested $500 B in 2015 on research. More than 65% was sponsored by commercial business and 28% by public funding through federal and state governments[2]. Commercial business funded technology development projects, whereas public funding was almost equally split between basic and applied science.

National Science Board[3]: Investment in R&D is a major driver of innovation, which builds on new knowledge and technologies, contributes to national competitiveness and furthers social welfare.

Several agencies try to justify their expenditure by measuring the impact and the value of the knowledge created by research in economic terms. The Organisation for Economic Co-operation and Development (OECD)[4] compares research activities and their economic impact across countries in its Science and Technology Indicators. The Global Innovation Index[5] is a similar composite indicator that ranks countries based on their innovation inputs and outputs. The Bloomberg Innovation Index[6] is a less extensive, but more understandable analysis of the 50 most innovative countries. The results from these analyses are communicated extensively in the press, largely for local self-promotion, but they do influence public perception and government policy. For example, the Conference Board of Canada[7] used a total of 21 indicators to measure innovation in 2015 for 16 countries. The Board defines innovation as "a process through which economic or social value is extracted from knowledge—through the creating, diffusing, and transforming of ideas—to produce new or improved products, services, processes, strategies, or capabilities". The response of the newly elected federal government of Canada to this analysis was to place a greater emphasis on science and discovery research, as reflected in the 2020 strategic plan for NSERC[8].

Society and its policy-makers are concerned about the ever-increasing cost of scientific research, the cost of new technology, and the cost of medical treatments. Most life science research in US[9] is funded by commercial companies, government grants and non-profit foundations. Almost 75% of clinical trials in the US are paid for by private companies. Large research projects are commonly funded by a mix of grants from various government agencies, institutions, and foundations, often internationally.

Federal spending in the USA on R&D is under budget pressure. Each year the federal budget proposes funding for Science, Technology, and Innovation with stated goals. The US has been dominant in global research across numerous industries, but China's R&D budgets are growing and are projected to surpass those of the US after 2022. Major US industries are projected to reduce investments in defense and aerospace R&D, but increase in energy, life science, information technology, chemicals and advanced materials. In the UK, the role of science and innovation in driving growth is recognized but annual UK government budgets[10] have been flatlined at £4.6 billion per annum since 2010.

Private companies fund research to develop, evaluate, deregulate and demonstrate the efficacy of their products. They consider the budgets for their research to be investments and select projects based on the potential return on their investment. Large US corporations apparently reduced their efforts in scientific knowledge creation between 1980 and 2007 as evidenced by a decline in scientific publications from company scientists, a decline in the market value of knowledge-based companies and a decline in the stock market premium that investors attach to scientific capability[11]. In contrast, the value attributable to technical knowledge has remained stable. Apparently, there has been a shift away from the creation of new knowledge (i.e. scientific publications) to the protection of existing knowledge for commercial application (i.e. patents).

Arora, Belenzon and Patacconi[12]: Large firms appear to value the golden eggs of science (as reflected in patents) but not the golden goose itself (the scientific capabilities).

They attributed this trend to increased global competition and narrowing of business scope. This trend may be a concern if large US companies now rely upon licenses or acquisition of startup companies to obtain new technology. Arora, Belenzon and Patacconi[13] raise the concern that startup companies often rely upon university research for their initial knowledge, which is subject to tightening budgets from government sponsors. If less knowledge is created by university research, the potential for innovation in large US companies may suffer. Alternatively, they point out that this trend may indicate a reallocation of research from inefficient large corporations to smaller, more efficient R&D organizations. Consequently, the improvements in the success and efficiency of research projects in these smaller R&D organizations may completely compensate for the quantitative reduction in large corporations[14].

Return on Investment (ROI)[15] is used by investors to estimate relative profitability of alternative investments. A positive ROI means higher net returns from sales compared to research and development costs. Jain, Triandis and Weick[16] cited several references indicating that the ROI from all scientific research is in the order of 30:1, but this varies among scientific disciplines and among industries. Most people and policy makers support the long-standing conclusion that investment by governments in research is overall cost effective.

Benjamin Franklin: An investment in knowledge pays the best interest.

ROI can be used to estimate the relative economic value of research projects into problems that cause a loss of value because that loss can be calculated with some accuracy. Solutions that add value to existing value chains or create innovations are more difficult because the estimates are only guesses about revenue. Consequently, ROI can compare the relative, but not the absolute, economic value of different research projects. ROI can make more accurate determinations in hindsight to determine the value of past research projects because the costs and revenues are more accurately estimated.

To determine the financial returns from a specific research project, commercial R&D organizations may use **Net Present Value** (NPV)[17]. This analysis estimates the value obtained based on an estimate of development cost and time. The NPV formula accounts for the fact that the investment in research will take several years before any sales of an innovation or a new product accrue. NPV is the total value of sales during a period of time minus research expenses adjusted for the "time value of money". NPV recognizes that time has an impact on value. A dollar tomorrow is worth less than a dollar today.

NPV provides a quick way to assess alternative business opportunities or products, but it can only estimate the relative financial returns from research, not the actual returns. The quality and reliability of the analysis depends on the quality of the assumptions and predictions estimated into the calculator. The closer the product is to market launch, the better the predictions. One limitation is that NPV assumes that research projects will be completed successfully, and that the knowledge will be implemented into a product. In other words, risk of failure is excluded from the calculation.

Lavallo and Kahneman[18] noted that executives routinely exaggerate the benefits and discount the costs of all types of projects, sometimes because they are overly optimistic. Therefore, NPV tends to be the most positive financial outcome possible from a research project. R&D organizations can be misled if the NPVs of potential research projects are compared using different assumptions and predictions. Despite its limitations, many businesses and commercial R&D organizations continue to use NPV or a similar tool to select the research projects that they will sponsor.

You can calculate NPV for your project based on the Investopedia calculator[19] (or similar calculator) to compare alternative research projects or research strategies based on a financial analysis. The discount rate in the NPV calculation will dramatically alter the NPV values[20]. It will be based on the relative risk tolerance of the sponsor. Investopedia[21] outlines some of the concerns about the use of Discount Rate and NPV for investing. Successfully introducing a product into the market-place depends on many factors beyond acquiring knowledge. NPV will fail to estimate a positive value for Observation, Modeling or Discovery projects because the duration from research to implementation is too long to make even guesses about revenue. NPV for these projects works only in hindsight. ROI is a better tool for those projects to give a relative indication of potential economic impact.

An example calculation will illustrate the difference between ROI and NPV. A Development project is conducted with an annual budget of $100,000 for three years, giving a total cost of the research project as

$300,000. It is anticipated that the product will be in the market for 10 years with a gradual introduction as it is accepted by consumers and a gradual decline as it is replaced by newer products. The total anticipated net revenue (after production and marketing costs) for the ten-year period is $600,000.

ROI is 100%, or 2:1, with a breakeven point in year 7. The initial investment is doubled over the 12 years with a net profit of $300,000. Assuming a discount rate of 10%, the NPV is $49,000. This means that the project is profitable and will return $49,000 considering the time-value of money. Note that NPV is affected dramatically by the discount rate and also by the duration of both the research phase and the duration of expected revenues. Delays during the research phase will reduce NPV and this is one reason why time is such a critical factor for research projects conducted in commercial R&D organizations. Use the Investopedia calculator or your own Excel table to determine the NPV and ROI of your research project and note the impact of delays.

Year	Cost	Revenue	Net Cash Flow	ROI	Discounted Cash Flow	NPV
0	$ 300	0	$ (300)		$ (300)	$ (300)
1	0	0	$ (300)		0	$ (300)
2	0	0	$ (300)		0	$ (300)
3		$ 20	$ (280)		$ 15	$ (285)
4		$ 40	$ (240)		$ 29	$ (256)
5		$ 60	$ (180)		$ 40	$ (216)
6		$ 80	$ (100)		$ 50	$ (166)
7		$ 100	$ 0	0%	$ 59	$ (107)
8		$ 100	$ 100	33%	$ 56	$ (52)
9		$ 80	$ 180	60%	$ 42	$ (10)
10		$ 60	$ 240	80%	$ 30	$ 20
11		$ 40	$ 280	93%	$ 19	$ 39
12		$ 20	$300	100%	$ 9	$ 49

Finding a sponsor for your research project is a specialized task that varies among geographic and political regions and among scientific disciplines. I cannot provide useful guidance to you on how to find an appropriate sponsor and defer to the experts in your R&D organization.

If you apply for a grant, the review committee will require more detail on the strategic and tactical plan for your research project than I have outlined here. If you receive the grant, you will likely have little subsequent interaction with the sponsor. Although the sponsor wants the research project to succeed, the sponsor who awards a grant rarely plays an active role in the project's management. Instead, these sponsors require financial reporting and often a report that includes some measure of success, such as a count of scientific publications or technical reports to customers. These reports may be annual, after designated milestones are achieved or at the end of the project.

If your sponsor is a commercial R&D organization, convincing the business leaders to invest will require detailed financial analysis beyond that outlined here. This analysis is only the beginning. The sponsor will also be your primary customer who will use the knowledge from your research to improve their business or create a new product. These sponsors will be more concerned with your planned accomplishments than with your strategy and tactical plans. These sponsors will be keenly interested in your success and will monitor progress closely, requiring frequent progress reports. An agile project management style that provides and seeks constant feedback with senior management is the best approach. The achievement of milestones on schedule will be essential. Also note that these sponsors have many opportunities and those opportunities are influenced by market conditions, which change constantly and sometimes quickly.

UNDERSTANDING CUSTOMERS

Your Envisioning, Modeling, Discovery and Development projects, like all other projects, have customers who will use the created materials, knowledge, discoveries, inventions and prototypes. These customers may be obscure, but they do exist. Customers differ from the sponsor. Often the customer is thought to be society because knowledge benefits everyone. But most research projects have specific customers. Addressing the needs of those customers to secure their support is as important in research as in any other project. Therefore, the project management manuals and team building manuals can provide guidance on how to manage your customers. Tool # 11 *Customer Assessment* lists some examples for you to consider as potential customers. Be inclusive at this stage and list all potential users beyond your primary customer.

If you are working as a member of a research team conducting a specific analysis as part of an experiment, you also have customers. Your customers are your colleagues on the research team who are expecting that you will deliver a product to them for their project/task/experiment. Interdependence is common in a research project with sequential tasks involving hand-offs, e.g. sampling > analysis > reporting. A virtual team where different tasks are done at different locations has an even higher degree of interdependence that requires mutual trust. For example, a greenhouse scientist who is imposing the treatments on the experimental plants and submitting samples for lab analysis should consider the lab analyst to be a customer. And in turn, the lab analyst should consider the statistician as a customer for the data. So, research projects of all sizes have customers who have expectations for quality and timelines. Who do you consider to be the customers for your research project?

The expectations that various customers may have for your research project, or in other words, **WHO wants WHAT by WHEN**, is essential to understand before committing to any research project. You should actively consult your potential customers when defining the objectives and deliverables, understanding the issues and revising your Research Concept from Tool # 9. You should constantly communicate with your customers before and during a research project, not only at the beginning and the end. This fundamental component of agile project management is essential for research project management. Ideally, the customer receives a series of "small wins" and provides feedback that aids in delivering the expected quantity and quality at the end of the project. These "small wins" may be peer-reviewed publications or presentations at scientific conferences or preliminary reports, but specific customers may require other types of reports. Iterative communication ensures that both the research team and the customer are constantly aligned and avoids "surprises" at the end of the project.

Research projects are different from other types of projects because they usually have multiple customers. Different customers may use the knowledge from your research project for different purposes - some of these purposes may even be in opposition. Your customers may have different requirements and different expectations of the deliverables. Conflicts will occur when an outcome is desired by one customer but considered to be harmful by another customer. For example, the scientific community expects full disclosure of all information on the performance and efficacy of a certain product. The investor (or the military) wants all information to be completely secret. The salesman wants only information that shows the product in a positive light, whereas consumers want a balanced objective report. Conflicts of this nature are common for industry-supported research projects at universities.

Another type of conflict may occur about quality. What is good enough for one customer may be inadequate for another. A customer sponsoring the research in a university for commercial purposes may conclude that sufficient information has been created based on one experiment (especially if it supports their marketing position), whereas the scientific journal that conducts peer review or the regulatory authorities may require additional experiments. The regulator assessing the safety of a product may have an even more stringent expectation of quality.

When drafting the Research Project Skinny using Tool # 9.4, it is critical to identify the primary customer for the research project. This is the customer whose needs must be satisfied for the project to be considered successful. Other customers will be considered secondary. It may be necessary to rank their importance to the success of the project. It may also be necessary to split the project into separate but interrelated projects that address the needs of specific customers independently. In Tool # 11 *Customer Assessment*, identify this primary customer and ensure that the assessment, expectations and deliverables described in the Research Project Skinny are consistent with those in Tool # 11.

Customers want you to achieve something specific in your research project, and in addition, they often want you to avoid doing something. Some things to avoid are obvious - safety, legal, ethical and regulatory violations cannot be tolerated. But there may be other outcomes that customers want to be avoid? For example, an entrepreneur may want to avoid any dependence on licenses for external intellectual property. A salesman may want to avoid any issues that would negatively impact sales of other products. An investor will want you to avoid public disclosure of information before filing intellectual property applications. Many research projects fail before the first experiment is even conducted because the researchers did not consider the expectations of their customers, or even that they had customers.

SELLING THE RESEARCH PROPOSAL

ONE OF YOUR most significant decisions that you must make is the selection of which research project to propose to a sponsor from your list of research concepts. Kevin Costner's character in the movie *Field of Dreams* believed that because he built it, they would come. Research is different. Just because you have an idea, the world will not beat a path to your door. You must sell your idea.

The Research Project Skinny from Tool #9.4 is a communication tool that you can use to motivate stakeholders, customers and the recruits to your research team. The Research Concept document from Tool # 9 is a more detailed outline that can be used in a formal research proposal to sponsors or investors as a request to provide funding. The communication to these different audiences must be simple, unambiguous, consistent and easily understood.

> *Naomi Oreskes: Our trust in science, like science itself, should be based on evidence, and that means that scientists have to become better communicators. They have to explain to us not just what they know but how they know it...*[1]

To select which Research Concept to take off your portfolio shelf, I suggest two different approaches. If the potential sponsor has clearly defined selection criteria that must be addressed, which is common for granting agencies, I suggest using a variation of Tool # 2 *Categorical Selections* that I have listed as Tool # 12.1 *Selecting Research Projects*. Every sponsor has a mandate and every research project that it sponsors must meet a set of defined criteria. Begin by understanding the selection criteria that the sponsor will use. These may be scientific, economic, training or even geographical. List these criteria as MUST HAVE and select only those projects that pass this selection (Tool # 2.1 *Pass-Fail Criteria*). Then create a list of WANTS and LIMITATIONS (Tool # 2.2 *Wants and Limitations Criteria*), or alternatively PLUSES and MINUSES, and rate each remaining project. The criteria used in your previous review of the Research Concept (Tool # 9.1 *SWOT Analysis* and Tool # 9.2 *Force Field Analysis*) may provide a useful set of criteria. Use your ratings to make a subjective selection or to make a prioritized list.

The alternative approach to prioritize a list of potential research projects is called Success Zones. This is the preferred option if the potential sponsor wishes to capture opportunities without predefined selection criteria, which is more common in a commercial R&D organization looking for ways to capture business opportunities. In his article, Peter Tinker[2] advises us to rate each potential project for probability of success as relatively high, medium or low. Then, rate the same potential projects for impact as relatively high, medium or low. For a research project, this might be scientific, societal or financial impact, depending on the objective from Tool # 8. The Success Zones approach can also be used to compare different visions, different scientific models or different strategies that address the same scientific problem. Note that Tool #12.2 *Success Zones* uses relative, not absolute ratings that use three categories. Tool # 2.5 *Pairwise Comparison Analysis* is one approach to make these relative ratings. Tool # 4.1 *Multivoting* is an alternative approach for a research team to prepare the ratings.

The high, medium and low ratings for each concept/project/strategy are entered into a 3x3 table showing success x impact. Once completed, the table helps you to make a final selection. Those that have low relative impact should be dropped from consideration regardless of their probability of success. You have better alternatives. Those with high impact and low probability of success are going to be a creative challenge. You will spend a lot of time planning their strategy and tactics. These are good concepts to keep on the back burner. Think about them, let them incubate and create more ideas. Your real great opportunity is the concept/project/strategy which has the relatively highest impact and relatively highest probability of success. This is the concept to propose to sponsors. The competition is likely to be intense because others have probably done a similar analysis, but nonetheless, this should be your top priority.

Many research sponsors require a final research proposal that contains a detailed strategy and a tactical plan for the experiments that will be conducted in your research projects. The use of research project management practices in the detailed planning of a research project is described in the next volume of the Research Project Management series *The Research Plan*. However, many research sponsors and decision-makers, especially those in commercial organizations, want to understand the "big picture" before committing resources to detailed planning. At this stage, you are ready to submit your application into this first round of consideration, seeking approval to proceed with the detailed planning. Even if you are an independent scientist in an academic institution with complete freedom to conduct any research project that interests you, you must convince yourself that further effort for detailed planning is warranted. Good luck with your proposal. But with careful preparation of your Research Concept in Tool # 9, and with consideration of your potential sponsors and customers in Tool # 12, you will not need it.

KEY POINTS

A CLEARLY DEFINED objective can be written for research programs, research projects, experiments and tasks. This statement empowers a research team to make decisions, provides focus on goals, identifies customers and helps stakeholders to align their expectations.

Tool # 8 *The Objective* provides guidance to define the desired outcome and the objectives of research programs, research projects, experiments and tasks.

An Outcome Statement can be used in grant applications and business proposals to explain the anticipated accomplishments of a research project.

A simple one-sentence Objective Statement for a research project is: **Our team will learn [WHAT].** Define "our team" with a distinct entity, group or individual, when known. Include **by [WHEN]** only if there is a fixed duration to the potential funding.

The concept for a research project is not to do something, but to accomplish something.

Tool # 9 *The Research Concept* guides you to prepare a document that summarizes the essential elements of a research project to be included in the proposal. The Research Concept includes the description of the scientific problem, your vision, the scientific model, including the knowledge gaps and its predictions, and the objective of one or more research projects that will address the knowledge gaps or test the predications.

The Research Concept is stored in a portfolio with your other concepts and must now compete for resources and financial support before strategic and tactical planning are initiated.

Tool # 9 *The Research Concept* presents three subjective analysis tools, which view and compare alternative concepts from different perspectives. Tool # 9.1 *SWOT Analysis* creates a subjective list of the Strengths and Weaknesses that are internal to the R&D organization and the Opportunities and Threats that are external. Tool # 9.2 *Force Field Analysis* recognizes the factors that are likely to help and those likely to hinder approval of the research project. Tool # 9.3 *Win Conditions* rates each potential research project for its ability to meet seven (or more) essential accomplishments.

The Research Project Skinny emphasizes brevity which forces clarity and enables effective communication. The Research Project Skinny states:

[WHO] will learn [WHAT] for [WHOM]. This research project will be considered finished when [WHAT] has been accomplished. This knowledge will [DO WHAT].

Sponsors fund scientific research. Their motive may be philanthropic to benefit society, but in most cases, they expect a return on their investment (ROI). Tool # 10 *Financial Analysis* guides you through ways to estimate the financial losses or gains associated with a problem, to prepare a preliminary budget and to estimate the potential financial returns.

Tool # 10.1 *Return on Investment* helps you estimate the potential ROI to a sponsor.

Tool # 10.2 *Net Present Value* is an alternative estimate of the potential returns to a sponsor considering the time-value of money. NPV accounts for the fact that the investment in research will take several years before any sales of an innovation or a new product accrue. NPV is the sum of the value of sales over a period of time minus expenses using the "time value of money" which recognizes that time has an impact on value. NPV is best used to determine the relative value of only Development projects.

Customers are the people who will use the knowledge created by your research project. Tool # 11 *Customer Assessment* is used to assess customers and to understand their expectations. Research projects, unlike other types of projects, usually have multiple customers, whose expectations may even be in opposition. Customer expectations can be stated as **WHO wants WHAT by WHEN**. This tool helps you to identify the primary customer whose needs must be satisfied and secondary customers who will also use the knowledge that you create.

Tool # 12 *Selecting Research Projects* guides you to select the best Research Concepts to submit to a sponsor for funding. Tool # 12.1 *Understanding Sponsors* can be used to make the selection if the potential sponsor has clearly defined selection criteria that must be addressed. This is more common in an academic R&D organization for applications to granting agencies. Tool # 12.2 *Success Zones* can be used to make the selection if the potential sponsor wishes to capture opportunities without predefined selection criteria. A determination of the probability of success and the importance of the outcome is a more common way used in commercial R&D organizations trying to capture business opportunities.

PART SIX

THE SCIENTIST'S TOOLBOX

OVERVIEW

MY EXPERIENCE IS in the management of research projects in the life sciences, but I believe that the principles and philosophy are universal. In this book series, *Research Project Management*, I provide research project management tools that have been useful to me and hopefully will be useful to you. I merge several concepts into this toolbox for the research scientist:

Creative thinking techniques ask critical scientific questions, identify problems and visualize solutions. Scientists need to think logically to create scientific models and to formulate testable hypotheses.

Leadership and teamwork empower a team of skilled, independently-minded scientists to cooperate, to coordinate, to make joint decisions and to act on those decisions.

The scientific method dictates an accepted philosophy to create knowledge. This requires scientific models that predict, hypotheses that can be tested and rigorous experiment designs that enable decisions.

Project management techniques from the traditional, extreme, adaptive and agile approaches enable planning, scheduling, monitoring and controlling of the experiments in a research project. A research project must be managed strategically and tactically to achieve its objectives.

Communication holds all of the other pieces together and determines whether a research project succeeds or fails.

These tools are written as recipes, step-by-step, much like the laboratory manuals that we are accustomed to using for analytical procedures. Creativity and decision-making tools are universal procedures that are used multiple times in a research project. Other tools are used in a sequential, interdependent process. All of the tools emphasize and enable communication. They are based on the premise that explicit short statements provide clarity and focus. Brevity eliminates jargon that often obscures understanding. Many of these tools will seem familiar to you, especially if you have conducted research projects previously. Some of these tools may contain tips to help you improve your existing management or participation in research projects. The tools are conceptual and by necessity they are generic, but I hope they provide you with a framework on which you can build your own processes and management style. Like all recipes, you will need to refine each tool for your organization's culture and your scientific discipline. I advise you, as a scientist, to experiment with using these tools in different situations, modify and improve them, keep what works and discard what does not work.

You can use research project management tools regardless of the size and complexity of your research project. Some of these tools may seem to be targeted to the Principal Investigator, but they are essential for all scientists in a research team. All of us plan and schedule our work. All of us work with others. Although the scale may be different, everyone on a research team can benefit from using these tools in their daily work. Even if your role is minor, you need to understand and support others to conceive and manage a research project. These are team tools. Therefore, explain these processes to your team and to your managers and to your administrators. To be successful, everyone needs to be working in the same direction with the similar expectations.

The principles and practices presented are meant to be exemplary, not prescriptive. These tools are intended to support a creative process. Not all R&D organizations have a mandate to be creative. We do not want our regulatory agencies to be creative in how they assess the safety and impact of the products that we use. We do not want medical researchers to be creative in how they treat their subjects in an experiment. In many cases, there must be strict guidelines on how research projects are conducted that will supersede the suggestions that I am making. Each R&D organization, each research project within that organization and each scientist within the research project will have different needs because they have different skills, different experience, different perspectives and different mandates. These tools will hopefully give you some options as you conduct your research project.

To improve these tools, I would be pleased to learn about your experience in research project management. Send comments and feedback to the author on The Scientist's Toolbox website at https://www.scientiststoolbox.com/.

TOOL # 1 BRAINSTORMING

WHAT:

A process of divergent thinking to collect ideas from a group of scientists without judgement, debate or criticism of those ideas.

Why:

To create as many ideas as possible.

To develop alternatives and options.

To collect different perspectives about a topic.

Input:

An open-ended scientific topic that has many alternative perspectives to consider.

A question that has many potential answers.

Output:

A long list of potential answers to the question.

How:

1) Organize a meeting. Clarify the objective by asking a specific question. The topic could be a business objective or a scientific question. For example: "How might we ...." or "How might [X] cause [Y]?"

2) Select your participants carefully so they reflect a diversity of opinion, technical expertise, perspective and culture. The number of participants will vary with the process but six to ten is optimal. Fewer participants limit diversity and more inhibits interaction.

3) Assign a Facilitator to lead the meeting and stimulate the discussion.

4) Facilitator assigns meeting roles to other participants: Time Keeper watches the time as allocated to each topic and informs the group. Scribe compiles ideas on a white board to facilitate discussion. Note Keeper records the ideas collected from the group. Process Checker ensures that agreed process is followed.

5) Prior to the meeting, provide the participants with reviews and other background information.

6) The participants conduct their own independent review of the topic. The ideas may be less spontaneous with this homework, but it is important to have a good technical understanding prior to a scientific discussion. Brainstorming on a scientific topic requires

technical expertise and knowledge. This is different from the more common brainstorming process that requires spontaneity.

7) Explain the process, duration and meeting norms. Emphasize that the process is to collect ideas, not to judge, debate or criticize.

First Round

8) Different brainstorming processes are applicable in different situations. Some will work better in one situation with one team and another in a different situation with another team. Customize the process to meet your needs and your R&D organization's culture.

9) Doing things differently creates discomfort and uncertainty but stimulates people to think differently.

10) Collect ideas using one or more of the following processes. Step-by-step details are described below for each. Combinations and variations of these processes are commonly used.

- Tool # 1.1 *Brainstorming Face-to-Face* is the most common style. Novel ideas are generated by the interaction among participants. Ideas can be explored in detail. The loudest voices may dominate the group. Introverted participants may not have the same opportunity as their extroverted colleagues to contribute.

- Tool # 1.2 *Brainstorming using Cards* is useful to collect the most obvious ideas quickly. Allows equal participation.

- Tool # 1.3 *Brain writing 6-3-5* is useful to collect the most obvious ideas quickly. Allows equal participation.

- Tool # 1.4 *Brainstorming using Bulletin Boards* provides time to think about details but there is a lack of interaction. Good for highly technical topics when detailed technical information, data and experiments need to be reviewed. Well suited for virtual teams.

- Tool # 1.5 *The Delphi Method* is a variation that is useful for virtual teams.

- Use Tool # 1.6 *Understanding the Topic* at any time during first or second round to create a better understanding of the topic, to organize the ideas and to summarize.

- Use any of the creative thinking techniques listed in Tool # 1.7 *Stimulating Creative Thinking* to stimulate discussion and create new perspectives. Encourage the participants to be imaginative and creative with their comments. Use different approaches for stimulation in different rounds of discussion.

Second Round

11) Reorganize and compile the ideas from the first round into themes for discussion in the second round.

12) Discuss each theme in sequence without criticism or debate. The participants ask questions for understanding and add more ideas, comments and revisions.

13) Identify similarities and differences among themes and among ideas within a theme.

14) Create new ideas by combining ideas within and between themes.

15) Add new ideas to expand on a theme or add new themes.

16) Compile all ideas for each theme and prepare a written summary.

17) End the meeting at the specified time. This is a one-time meeting. There is no follow-up meeting. There are no decisions made.

18) Note Keeper distributes the written summary to participants. Avoid attributing ideas or comments to individuals in the notes because this may inhibit idea creation. Archive the written summary for future reference.

19) Decide on follow-up actions and next steps. Initiate a convergent thinking process to make selections from the brainstorming list; for example, use Tool # 2 *Categorical Selections*, Tool # 3 *Review Panel* or another convergent thinking process.

TOOL # 1.1 FACE-TO-FACE

a) Ask each participant to contribute their ideas and thoughts in a sequential roundtable process.

b) Keep everyone motivated and enthusiastic by having several rounds of discussion. For example, each participant may be restricted to contribute only one idea per round.

c) Avoid criticism or debate. Allow only questions for clarification and expansion of ideas.

d) Continue until all ideas are collected and proceed to round two (step 11).

TOOL # 1.2 CARDS

a) Ask each participant to write one and only one idea on a paper card.

Then, pool the cards centrally.

b) Ask each participant to pull a card, randomly and then to comment on the original idea, improve the original idea or add a new idea. This is done silently with no group discussion and the card is returned to the pool.

c) Repeat this process for a specific number of rounds.

d) Continue until all ideas are collected and proceed to round two (step 11).

TOOL # 1.3 BRAIN WRITING 6-3-5

a) Invite six people. Give each a form that has a table of six rows (one for each participant) and three columns (one for each idea).

b) Ask each participant to fill the first row of the table with three ideas. Limit the time to 5 minutes. Avoid speaking or commenting.

c) Pass the form to the next person on their right. Ask each participant to fill the second row trying to improve or expand on the three ideas in the first row.

d) Continue this passing and commenting process for as many rounds as there are participants (usually six). At the end, every participant has added a comment to each of the original 18 ideas (six participants contributed three ideas each). Proceed to round two (step 11).

Variations of Brain writing 6-3-5:

The number of columns can vary, but 3 is often used.

The number of participants can also vary. The process can be modified to facilitate input from a large group of participants by passing the forms only a specific number of times. This allows more people to contribute more than 18 ideas, but it may limit the depth.

The Brain writing 6-3-5 meeting can be held as a virtual meeting. Ideas are circulated in sequence by email or by updating forms on a server.

TOOL # 1.4 BULLETIN BOARDS

The participants meet over the internet. Both public and proprietary online service providers are available to host discussions. Consider how much security and confidentiality is required when choosing a service provider. An electronic bulletin board allows participants at different locations to contribute. Be careful with intellectual property. Premature disclosure of ideas on a public network may be viewed as public disclosure and may negate claims to intellectual property.

a) Organize the bulletin board. Post the background information, literature reviews and technical information relevant to the topic.

b) Ask the participants to review this information and add their own ideas and comments on the board.

c) Set a time limit when the board will be closed. The board may be active for several days or weeks. This allows participants time to think about their ideas and to gather technical information if needed and the same time, share their ideas with others. Be careful not to include judgement, assessment and selection because this will inhibit idea creation.

d) Stimulate discussion by asking questions, perhaps in multiple rounds. Use one or more of the creative thinking processes listed in Tool # 1.7 Stimulating Creative Thinking.

e) At the end of the first stage, reorganize the ideas into themes and initiate a follow-up discussion of each theme in a second round. Use Tool # 1.6 Understanding the Topic to help organize and summarize the ideas.

f) The session is closed on a specific date.

TOOL # 1.5 THE DELPHI METHOD

The Delphi Method is an iterative process that uses a series of questionnaires to generate written discussion and argument. One version is described in Tool # 3.2 *The Delphi Method* to obtain feedback from geographically dispersed experts. A modification of the Delphi Method is an alternative style brainstorming that allows you to obtain input from technical experts without requiring them to meet face-to-face. This is a useful tool for virtual teams that are widely dispersed across geographical locations, but it is time consuming. Participants and responses may or may not be anonymous.

a) This iterative process uses a series of questionnaires to generate anonymous written discussion and argument and is very demanding of the Facilitator role.

b) Facilitator creates questionnaires, distributes them electronically and compiles the responses. Experts provide answers. Each iterative round of questions and responses builds on the results of the previous one. The internet makes the distribution of questions and collection of responses relatively simple, but nonetheless, this is a time-consuming process and will require a significant commitment. A bulletin board (Tool # 1.4) is an alternative method to collect responses.

c) Invite the participants. Usually 6 to 10 participants are invited to participate. By agreeing, the participants are committing a specific amount of time on a specific schedule.

d) Prior to the meeting, the Facilitator provides the participants with all relevant information.

e) Outline the process and seek agreement from the participants. For example, the participants must respond in a timely manner, otherwise the process becomes too protracted and ineffective.

f) The Facilitator prepares and distributes the first questionnaire. Ask the participants to review this information. Ask participants to add their own ideas and comments. The Facilitator compiles and distributes the responses.

g) The Facilitator prepares and distributes the second questionnaire asking specific follow-up questions. Any of the approaches in Tools # 1.6 and # 1.7 may be appropriate to understand the responses from the first questionnaire or to stimulate discussion from a different perspective.

h) The Facilitator prepares and distributes the third and subsequent questionnaires as needed to gather more ideas or to clarify ideas.

i) Continue for as many rounds as required to gather all ideas.

j) Go to the second round at step 16 to complete the session.

TOOL # 1.6 UNDERSTANDING THE TOPIC

When you want to create a better understanding of the topic, you can use one or more of these approaches. See the text or online references for detailed descriptions of each.

- Repeated Questions allows you to dig deeper and think laterally about the topic. Ask the group, one of these questions repeatedly: Why? How? Why else? How else? What is stopping us?

- Is/Is-not Matrix is a simple two column table, indicating what the topic is and what it is not, to better understand a problem, opportunity or scientific question

- Four Windows is a summary diagram prepared to better understand how two levels of two factors interact.
- Mindmap provides a visual depiction and organization of related ideas into themes.
- Ishikawa or Fishbone diagrams (Tool # 7.3) visually depict cause-and-effect relationships when a single effect may have multiple causes. This graphical method seeks to identify the root causes of a specific event.
- Interrelationship diagrams (Tool # 7.4) visually depict potential cause-effect relationships when multiple effects are created by multiple causes.

TOOL # 1.7 STIMULATING CREATIVE THINKING

When you want to stimulate the group to think differently or creatively about the topic in any of the above approaches, you can use one or more of these approaches. See the text and references in the Further Reading section for detailed descriptions of each.

- Use pictures instead of words to communicate ideas. Drawings, pictures, graphs and symbols can often be used to represent complex ideas more clearly.
- List and challenge the assumptions in rival scientific models or hypotheses. How are they similar? How do they differ?
- Reword the original question using synonyms or different words.
- Identify undesirable or negative outcomes. How can we make this fail? What is the worst-case scenario? Then reverse these actions and do the opposite to define a positive response. This can be a playful or humorous exercise.
- Dissect a complex problem into its relevant attributes including technical, scientific or business attributes to make the topic easier to discuss.
- Use role playing to view the topic from different perspectives using De Bono's Six Thinking Hats. For scientific topics it can be useful to use different hats than what De Bono suggests. For example, in the life sciences, you may wish to view a topic from the molecular, biochemical, cellular, genetic, organismic and ecological points of view. A variation would be to view the topic

from the perspective of different scientific models that make different predictions and hypotheses. Alternatively, you may wish to view a topic from the perspective of different customers or different stakeholders.

- Transform an idea using SCAMPER.
- Use analogous reasoning. Adapt ideas from similar scientific phenomena or related scientific disciplines. Has a similar situation existed in the past, in another project, in another scientific disciple? What did those people do?
- Use biomimicry. Does a similar problem exist in nature? If so, how has nature solved a similar problem?
- Combine elements of existing ideas in a new way using the Heuristic Ideation Technique.

TOOL # 2 CATEGORICAL SELECTIONS

WHAT:

A process of convergent thinking that analyzes a set of options to select the best in a given situation considering multiple, potentially conflicting criteria.

Why:

To make selection decisions among ideas, problems, objectives, risks, hypotheses, experimental designs and many other choices in a research project.

Input:

A list of ideas, options or alternatives.

A list of ideas from Tool # 1 *Brainstorming*.

Output:

One of the following types of selection decisions:

A: Pass a threshold to qualify.

B: Ranking from high to low.

C: Best, "winner takes all".

How:

1) Review the list of the options or ideas, for example the outcome from Tool # 1 *Brainstorming*.

2) State the type of selection decision that needs to be made: Type A, B or C selection.

3) Review the background information. A thorough technical knowledge of the topic may be required.

4) List the criteria that should be considered in the selection process. These may be strategic, technical, scientific and/or business criteria.

5) Different selection processes are applicable to different types of selections. Some will work better in one situation with one team and another in a different situation with another team. Customize the process to meet your needs and your R&D organization's culture.

6) To make a Type A decision - pass a threshold to qualify:

- Use Tool # 2.1 *Pass-Fail Criteria* to shorten a long list of options into a shorter more manageable list for further selection.

- Use Tool # 2.3 *Pareto Analysis* to identify the options or ideas that

The Research Proposal

have the greatest impact.

- Use Tool # 4.1 *Multivoting* as a team to shorten a long list of options based on each individual's assessment.

7) To make type B (ranking) or type C (winner-take-all) selections from a short list, compare the rankings based on the scores from a subjective or objective analysis tool. Use a scale that compares one option relative to another.

- Use Tool # 2.2 *Wants and Limitations Criteria* to consider the positive and negative aspects of an option or idea.
- Use Tool # 2.4 *Decision Matrix* to compare attributes of each option if a categorical Yes/No rating of the criteria is insufficient to compare options.
- Use Tool # 2.5 *Pairwise Comparison Analysis* to compare one option relative to another, either as an entity or for each criterion.
- Use Tool # 4.1 *Multivoting* as a team to rank options.
- Use Tool # 4.2 *Consensus* as a team to make a final selection that everyone on the team can support.

8) Avoid being too numerical for your final selection. Let the mathematics in each tool be only a guide. Use your judgement to review the results and make the final selection. Modify the criteria to determine how this impacts the selection. Consider the impact and consequences of the selection – both positive and negative.

TOOL # 2.1 PASS-FAIL CRITERIA

1) Select the criteria that are essential Must-Have criteria.
2) Evaluate each option for the Must-Haves using a categorical rating – Yes or No.
3) Any option that fails to meet all of the Must-Have criteria is eliminated.
4) For type A decisions (threshold to qualify) this selection may be sufficient.
5) For type B and C decisions proceed to one or more of the other tools below, or to Tool # 3 Review Panel, or to Tool # 4 Group Voting.

TOOL # 2.2 WANTS AND LIMITATIONS CRITERIA

1) List the nice-to-have or Want criteria that represent desired positive aspects. These are considered less important than the Must-Have criteria used in Tool # 2.1 but are desirable benefits to be captured.

2) List the Limitation criteria. Include factors that constrain the scope or quality of the project. Include potential adverse outcomes, threats or risks that should be avoided.

3) To keep it simple, assume that each criterion has the same importance, impact or significance. If this is not the case, use Tool # 2.4 Decision Matrix with weighted criteria to make the selection.

4) Evaluate each option for each Want criterion using a categorical rating. Is this captured? – Yes or No.

5) Evaluate each option for each Limitation criterion using a categorical rating. Is this avoided? – Yes or No.

6) Total the number of Yes ratings to determine a final score and rank the options from high to low.

TOOL # 2.3 PARETO ANALYSIS

1) Use this as an alternative to Tool # 2.1 Pass-Fail Criteria for type A threshold selection decisions or to make type B ranking decisions.

2) List the potential causes of an effect in the first column of a spreadsheet. Alternatively, this might be potential risk events or any set of alternatives.

3) Assign a quantitative measure to each in the second column. This may be the magnitude of the effect, the probability of occurrence or frequency of occurrence.

4) Sum the quantitative measurements for all options.

5) Calculate the relative magnitude/frequency of each option as a percentage of the total. Enter this value in column 3.

6) Rank the options from high to low using the values in column 3.

7) Calculate the cumulative frequency starting with highest and continuing to the lowest using the values in column 4.

8) Select those options that contribute to the 80% cumulative sum. Discard those that provide the remaining 20%.

Example of Pareto Analysis:

The effect and relative effect of six factors.

Criterion	Effect	Relative Effect
A	70	70/297 = 24%
B	20	20/297 = 7%
C	85	85/297 = 29%
D	27	27/297 = 9%
E	53	53/297 = 18%
F	42	42/297 = 14%
Total	297	297/297 = 100%

The cumulative effect of six factors. Those factors contributing 80% of the relative effect (Criteria C A E and F) are selected. Those contributing the remaining 20% (Criteria D & B) are ignored.

Criterion	Effect	Relative Effect	Cumulative Effect
C	85	29%	29%
A	70	24%	52%
E	53	18%	70%
F	42	14%	84%
D	27	9%	93%
B	20	7%	100%
Total	297	100%	

TOOL # 2.4 DECISION MATRIX

1) List the nice-to-have or Want criteria that represent potential positive aspects.
2) List the Limitation criteria that represent potential adverse aspects.
3) Rate each option on a 1-10 scale based of how well it benefits the idea or option.

4) Optional: include a weighting factor for each criterion based on its relative importance.

5) Calculate a total score for each option by summing the criteria scores.

TOOL # 2.5 PAIRWISE COMPARISON ANALYSIS

1) Make a PCA table by listing the options from the Pass-Fail selection (Tool # 2.1) in the first column of table. Repeat the list as the headings for each column.

2) Compare the options pairwise cell-by-cell in the table. In each cell identify the best option using an overall comparison.

3) Mark the best option in the cell.

4) Total the number of positive comparisons for each option and rank the options based on the score.

5) To use the Analytic Hierarchy Process, compare options using specific Want or Limitation criteria and total the score. Repeat the PCA for all criteria and calculate a total score.

TOOL # 3 REVIEW PANEL

WHAT:

A process to obtain feedback from others.

Why:

To select among several options when multiple, potentially conflicting technical criteria need to be considered by scientific experts.

To select among several options when topics are highly important, highly technical or highly political.

Input:

A short list of options or alternatives, for example from Tool # 2 *Categorical Selections* or from Tool # 4.1 *Multivoting*..

Ideas, visions, scientific models or concepts concerning a scientific phenomenon.

Questions and options requiring advice from experts.

Technical information, analysis or issue statements about a scientific phenomenon.

Output:

A deeper understanding of a scientific phenomenon.

The answer to a specific question.

Identification of issues, tensions and concerns about an idea or concept.

Feedback to improve an idea or concept.

How:

Use Tool # 3.1 *Expert Panel* to conduct a detailed technical evaluation of alternative options by experts who can meet face-to-face.

Use Tool # 3.2 *The Delphi Method* to conduct a detailed technical evaluation of alternative options by experts at multiple locations who are unable to travel.

Use Tool # 3.3 *Deliberative Forum* to conduct a discussion about issues and options within the research team, with a group of informed stakeholders or with an interested panel.

TOOL # 3.1 EXPERT PANEL

1) Assign the role of Facilitator.

2) Ask one or more specific questions that you want a group of technical experts to address, such as "Which of these three options is the best way to achieve ..... ?" Avoid open-ended questions that lead to brainstorming.

3) Select a group of participants based on their diversity of technical knowledge, opinions and perspectives relevant to the topic. Usually 3 to 5 participants are invited to participate.

4) Prior to the meeting, Facilitator provides the participants with all relevant information, for example, literature reviews and summaries of the brainstorming and selection meetings.

5) Assign meeting roles to other members of the research team, not to the experts. Time Keeper watches the time as allocated to each topic and informs the group. Scribe compiles ideas on a white board to facilitate discussion. Note Keeper records the ideas collected from the group. Process Checker ensures that agreed process is followed.

6) Facilitator summarizes the desired outcome of the selection process and the alternatives that have been prioritized so far.

7) Use a structured sequence for the meeting that enables information sharing, discussion and decision-making.

8) To gather information and opinions, use a roundtable process in which each participant presents their findings. Ensure that the emotional issues influencing beliefs, biases and perspectives are understood. Ensure that all facts, ideas, options and perspectives are understood. No criticism or debate is allowed at this stage - only questions for clarification and expansion of ideas.

9) The participants discuss the topic until all facts and technical details are collected and understood.

10) Once the facts are collected, conduct a second round of open discussion. The participants may debate the scientific topic and may challenge one another. Keep the debate professional. Avoid criticism of beliefs, experience or perspective because this can become personal and lead to conflict.

11) The output from the meeting should be a specific recommendation to answer the question posed to the group. To reach a conclusion, use a consensus voting process to make a final recommendation (Tool # 4.2 *Consensus*). The output is not a binding decision for the research team.

12) Usually, an expert panel is a one-time meeting. In exceptional cases

when the topic is highly technical, iterative information gathering, sharing and discussion may require additional meetings.

TOOL # 3.2 THE DELPHI METHOD

1) Assign the role of Facilitator. This iterative process uses a series of questionnaires to generate written discussion and argument and is very demanding of the facilitator role. The responses may or may not be anonymous. Keep responses anonymous if this will reduce conflict or increase participation.

2) Facilitator creates questionnaires, distributes them electronically and compiles the responses. Experts provide answers. Each iterative round of questions and responses builds on the results of the previous one. The internet makes the distribution of questions and collection of responses relatively simple, but nonetheless, this is a time-consuming process and will require a significant commitment.

3) Invite technical experts as participants. Usually 6 to 10 participants are invited to participate. This format allows more "experts" to participate than Tool #3.1 *Expert Panel* but the work of the Facilitator increases with each additional participant. By agreeing, the participants are committing a specific amount of time on a specific schedule.

4) Prior to the meeting, the Facilitator provides the participants with all relevant information, literature reviews and the summaries of the brainstorming and selection meetings.

5) The Facilitator outlines the process and seeks agreement from the participants. For example, the participants must respond in a timely manner, otherwise the process becomes too protracted and ineffective.

6) The Facilitator prepares and distributes the first questionnaire. Begin by asking one or more specific questions, such as "Which of these three options is the best way to achieve ..... ?" Avoid open-ended questions that lead to brainstorming. The Facilitator compiles and distributes all the responses.

7) The Facilitator prepares and distributes the second questionnaire. This includes a summary of all responses from the first questionnaire, anonymously, and asks the participants to respond to the summary. It is likely that many of the responses will be critical. The Facilitator compiles and distributes the responses from the

second questionnaire.

8) The Facilitator prepares and distributes the third questionnaire. This includes all responses from the second questionnaire and asks the participants to prioritize the options or to make a selection and provide their reasoning. The questions may be worded in a way that enables the participants to vote for their preference.

9) The Facilitator repeats this cycle of questions, responses and voting until a consensus is reached, or all opinions are fully vetted. Usually three to five rounds are sufficient. The Facilitator strives to guide a group towards a consensus decision and words the questions accordingly.

10) The report from the Facilitator should be a specific recommendation to answer the question posed to the group.

TOOL # 3.3 DELIBERATIVE FORUM

1) Assign the role of Moderator.

2) State a specific contentious issue that you want a group of informed and interested individuals to consider. Ask a specific question, such as "Which of these options is the best way to achieve ..... ?"

3) Select a group of participants based on expertise, experience, thinking style and beliefs. The participants should be informed, creative, collegial but think independently. Usually 10 or more participants are invited.

4) Prior to the meeting, Moderator provides the participants with an issue guide that outlines options to be considered with both positive and negative arguments. These may be documents concerning your scientific problem (Tool # 5), your vision of the future (Tool # 6), scientific models (Tool # 7), objectives (Tool # 8), research concepts (Tool # 9) or other issues about a scientific phenomenon.

5) Assign meeting roles to other members of the research team. Time Keeper watches the time as allocated to each topic and informs the group. Scribe compiles ideas on a white board to facilitate discussion. Note Keeper records the ideas collected from the group. Process Checker ensures that agreed process is followed.

6) In a round table format, ask each participant to share their feelings, concerns and initial response to the question. Ensure that the emotional issues influencing beliefs, biases and perspectives are

understood.

7) Review each option individually as described in the guide to understand its positive and negative aspects.
8) Identify the tensions or opposing viewpoints. Identify agreements and consensus viewpoints. Avoid debate and conflict. Identify the options which might be improved to resolve some of the tensions.
9) In the final phase of the discussion, prepare an answer to the question including the opposing viewpoints. The forum does not make a formal decision, only a recommendation.
10) Note Keeper prepares a report that is sent to the research team or Principal Investigator.
11) Usually a forum is a one-time meeting.

Notes:

- Prepare an informative, unbiased guide that outlines options with both positive and negative arguments. The documents and descriptions concerning your scientific problem, your vision of the future and the scientific models are well suited to this review. Add your assessment of the constraints, wants, limitations, risks and threats posed of different options.
- Select your participants carefully. You want diversity of expertise, experience, thinking style and beliefs to fully vet the options. The participants should be informed, but not experts; they should be advocates, but not argumentative; they should be creative, but pragmatic; they should be collegial, but think independently. The discussion should be open and time limited.
- An option is to assign roles, such as devils' advocate or angel's advocates to individuals or groups. Use this option if opposing viewpoints are not expressed. Be transparent in this assignment.
- An option is to assign perspectives to individuals, such as different customers or different stakeholders, if they are not present.
- An option is to use De Bono's Six Thinking Hats process using the original hats, or using other perspectives, including scientific perspectives, such as ecological, genetic and biochemical.

TOOL # 4 GROUP VOTING

WHAT:

A voting process for a team to make a selection decision.

Why:

To make a team-based selection decision considering the type of selection needed, the importance of the selection and the need for team support to implement the decision.

Input:

A list of options.

Output:

One of the following types of selection decisions:

A: Pass a threshold to qualify.

B: Ranking from high to low.

C: Best, "winner takes all".

How:

1) Usually group decisions are made at a face-to-face meeting.

2) Assign meeting roles. Time Keeper watches the time as allocated to each topic and informs the group. Scribe compiles ideas on a white board to facilitate discussion. Note Keeper records the ideas collected from the group. Process Checker ensures that agreed process is followed.

3) Frame the issue so everyone understands the decision needed and why it is important.

4) Scribe writes the list of the options to be considered on a board.

5) Facilitator asks each individual to disclose their beliefs, feelings, concerns and prior experience on the topic.

6) The participants discuss any emotional, ethical, moral, cultural or political issues that may impact the decision or acceptance of a decision by stakeholders.

7) The participants discuss the technical and scientific topic.

8) Choose one of the following voting methods to make the selection decision.

- Use Tool # 4.1 *Multivoting* to create a short list of options or to make Type B selections (ranking). Each of the selected options has the support of one or more members of the group.

- Use Majority Rule if a quick decision is needed and any decision will be accepted by the group.
- Use Tool # 4.2. *Consensus* to make a decision that everyone can support. Use this style when acceptance by the group is important to implement the decision. This takes time - the longer the list or options, the more time required.
- Use Agreement if everyone must agree unanimously on the selection or decision. Check these types of selection decisions carefully for symptoms of Groupthink.

TOOL # 4.1 MULTIVOTING

1) Inform the group on how many options must be on the final reduced list.

2) Each participant may use any subjective or objective analysis tool that they wish for their individual decision-making, for example Tool # 2 *Categorical Selections*.

3) The Facilitator decides how many votes each member will have based on the size of the initial list. A rule of thumb is that each participant gets a number of votes equal to half the number of options. For example, if there are 10 options, each member gets 5 votes.

4) The Facilitator decides how these votes may be distributed among the options. There are many alternatives, such as:

Secret ballots. Votes are written on paper anonymously.

Limit to one vote per option. The participant cannot place two votes on one option.

No Limits on the number of votes per option. If a participant feels strongly about an option, all votes may be cast for that option, or different options may get one vote each, or any combination to a maximum number of votes.

Each member ranks their choices in order of priority, with the first choice ranking highest.

5) The Facilitator tallies the votes. If a Type B selection decision is clear, stop. Proceed if there is need to create a short list with a specific

number of options, for example, to send to Tool # 3 *Review Panel*.

6) Any option with no support is eliminated.
7) Each participant explains their selections. In an open discussion, clarify any incorrect information or misunderstandings. Improve any ideas. Discuss feelings and emotions. Discuss biases and beliefs based on experience. Discuss assumptions, facts and logical arguments.
8) Repeat the voting for a reduced number of options using a reduced number of votes per individual to select the next set of top options.
9) Continue discussing and voting until the required number of options, or the required ranking is completed.
10) Proceed to Tool # 4.2 *Consensus* if a type C selection decision is needed.

Notes:

- Multivoting is a fast and simple way to eliminate options with no support.
- Multivoting is preferable to straight voting because it allows an item that was weakly supported initially to rise to the top.
- Individuals may use any of the steps in Tool # 2 *Categorical Selections* to make their individual choices.
- A disadvantage of multivoting is that embryonic ideas may be eliminated too quickly. To avoid this problem, give time for discussion, brainstorming and revision of the options.
- Avoid peer pressure to conform.
- Ensure adequate discussion and consideration of options.
- Sticky dots are good for voting to ensure that everyone casts the same number of votes. They can also be easily removed from the board during multiple rounds of voting. Sticky notes can be used to include comments with the votes or to indicate priority.
- If a decision requires everyone's agreement, continue discussing and voting until the selection decision or the required ranking is unanimously agreed. There is a strong tendency to promote Groupthink with this approach because some people just become tired and want to go home without really supporting the selection.

TOOL # 4.2 CONSENSUS

1) The consensus process works best on a short list of about 3-5 options because each option is discussed in detail from several perspectives. Tool # 4.1 *Multivoting* works well to establish a prioritized list.

2) The group discusses the options before voting to clarify and to determine if any can be improved. A second round of voting is needed to reprioritize the options after a discussion.

3) The Facilitator asks each individual whether they can more-or-less support the first option. This can be done by blind voting to ensure that dissenting opinions are captured.

4) If everyone can support the first option, the discussion stops.

5) If any individual cannot support the first option, that option is discarded.

6) The Facilitator then asks each individual whether they can more-or-less support the second option.

7) This voting on the prioritized list of options continues until everyone can more-or-less support one of the options.

8) If the group cannot make a consensus decision (i.e. none of the options can be supported by everyone), the Principal Investigator or Project Manager or another individual makes an autocratic decision.

Notes:

- Support in consensus voting is an ambiguous concept. Some project management manuals require at least 70% support, but that is a poorly defined measurement. More-or-less support is a vague term that has different meaning to different people.

- This voting process works only if all participants openly express their opinions and disagreements.

- Avoid intentionally or unintentionally creating peer pressure to conform. A role for the Process Checker in the meeting is to prevent intimidation or personal conflict during the discussion. Participants must be free to think independently and critically.

- Be careful because Groupthink may develop with multivoting and is almost certain to be occurring if unanimous decisions are common.

TOOL # 5 THE PROBLEM STATEMENT

WHAT:

A description of a scientific problem.

An understanding of the current situation and the justification for a change.

Why:

To create clarity and focus for a research project.

To understand a problem that requires new knowledge, adaptation, invention or innovation.

To communicate a need to others.

To motivate and empower your research team, stakeholders and sponsors.

Input:

A problem that requires a solution to benefit people or a business.

A recognized or unrecognized need.

A scientific challenge or curiosity.

Output:

Problem Description to be included in business proposals and in grant applications.

A one-sentence Problem Statement that will be used in communications: **[WHO] needs [WHAT] because [WHY].**

How:

1) Select a problem that merits attention based on scientific, commercial or societal need. The problem may be commonly recognized in society, an unrecognized need for a product, a business opportunity in a new market, a needed process improvement, a need for regulatory guidelines or a scientific question sparked by curiosity.

2) Frame the problem in different ways. Frame the problem as a loss by focusing on the negative consequences. Frame the problem as a potential gain by focusing on the positive opportunities. Use different words to express the problem. Use synonyms in your description. Reverse the problem or assumptions. View the problem from different people's perspectives.

3) Observe and describe the problem. If possible, interview people

experiencing the problem first hand. Who is involved or impacted by the problem? When does the problem occur? Where does the problem occur? Why does the problem occur? How does the problem develop?

4) Gather the facts. Conduct a review of the scientific literature. Write a general overview of the current scientific knowledge on the problem. Avoid technical detail at this early stage. Collect quantitative data, specific facts, numbers and other details. Add visual observations, photographs, video or qualitative data.

5) As questions. What scientific topics are related to the problem? What are the current scientific models?Are the current scientific models adequate? What assumptions have been made? Do they appear to be valid? What knowledge is missing?

6) Break a complex problem into simpler attributes - characteristics, components or scientific topics. Relevant attributes may include technical, scientific or business characteristics. In life sciences, six scientific attributes might be physiology, genetics, ecology, biochemistry, molecular biology and biophysics. Ask "How does each attribute contribute to the problem?" Ask "Which attributes are the major drivers creating the problem?" Ask "Do the attributes have a relationship with each other?"

7) Delve deep into the root cause of the problem. Symptoms can be misleading. Ensure that you are addressing the real root cause of the problem. Ask repeated questions about each of the facts. Ask Why? or How?. Then after each answer, ask Why? or How? again. Keep asking the question repeatedly until the root cause is identified. Ask Why/How Else? to think laterally and add more causes.

8) To understand the consequences from each of the facts. Ask: So what? What are the implications of that fact? Why is this fact important?

9) Use a visualization tool, such as mind mapping or cause-effect diagram (Tools # 7.3 and 7.4), to visualize relationships and to group the causes into categories or themes.

10) Maintain a broad perspective. Consider all options and remain open to fresh ideas. Do not delve TOO deeply into the scientific detail yet. Maintain a general overview on the problem.

11) Classify the problem into one of the following four categories:

    A.   Problems that are poorly described.

B. Problems that lack understanding.

C. Problems that have several hypothetical solutions.

D. Problems that need validation of a solution.

12) Write a draft Problem Description that includes the current knowledge, the technical details, your observations and your analysis. Describe the problem's attributes, its effects and its putative root cause(s) based on your analysis. Indicate why the problem is important to solve. This summary may be used in grant applications or business proposals.

13) Obtain feedback from others and revise your Problem Description. Discuss the problem with others to gather different perspectives using Tool # 3.3 Deliberative Forum.

14) Write a one sentence Problem Statement: [WHO] needs [WHAT] because [WHY]. Brevity forces clarity and simplicity. Brevity provides focus. Brevity enables effective communication. Name a specific customer as [WHO]. This is a tentative guess. The customer may change later depending on their need and level of support. Write multiple Problem Statements if the original problem can be split into distinct, separate attributes or if multiple customers have different needs. Reword the problem to change your focus and perspective. Then, select the "best" wording.

15) Next step (optional): If you are comparing several problems for opportunities, use Tool # 2 *Categorical Selections* to select problems that merit your further attention. Avoid discarding any problem; keep problems for possible research projects in a notebook. Repeat periodically to obtain new information for your analysis. Opportunities may arise that will change your prioritization and selection.

16) Next step (optional): Do a more in-depth analysis of commercial opportunities. Many commercial R&D organizations conduct a full Observation project to understand a problem before they commit to commercial development. This may include legal, market and value chain research to determine how to capture value and what their potential return on investment might be. If a problem is poorly described (Type A problems), an Observation project may be needed to increase your knowledge about the scientific phenomenon contributing to the problem.

Notes:

- Discovering problems requires just as much creativity as discovering solutions (according to Einstein it requires more). Only by explicitly stating the problem, can you develop a solution. This applies to all types of problems, regardless of their complexity.
- How the Problem Statement is worded skews your subsequent vision and the objectives of your research project towards acquiring specific knowledge.
- The Problem Statement can be applied to research programs, to Observation, Modelling, Discovery or Development projects, or to experiments - in other words, to any problem that requires new knowledge. There is a difference in scope, not process.

TOOL # 6 THE VISION STATEMENT

WHAT:

A vision of what the future might be if a solution is implemented.

An opportunity that can be captured by changing the current situation.

A vision for a product or a technology that will create an innovation.

A description of the potential consequences of a solution.

Why:

To envision the best possible solution to a problem.

To motivate your research team, customers, stakeholders and sponsors.

To empower your research team.

Input:

The Problem Statement from Tool # 5: [WHO] needs [WHAT] because [WHY].

The Problem Description from Tool # 5.

Output:

A Vision Description that will be used in business proposals and in grant applications.

A one-sentence Vision Statement that will be used in communications: **If [CUSTOMER] is able to [DO WHAT] then [SOLUTION].**

How:

1) Review the Problem Statement and Problem Description from Tool # 5.

2) Reword the loss or negative consequences in the Problem Statement into an opportunity to increase or improve something.

3) Use Tool # 1 *Brainstorming* to search for all possible solutions, adaptive or radical, that will capture the opportunity. You may use any divergent thinking tool either as a team or as an individual. List all options and ideas. Be thorough and consider all possibilities. Collect all ideas BEFORE you judge, criticize or discard any.

4) Organize the ideas into categories and themes. Use Tool # 1.6 *Understanding the Topic* to help organize and summarize the ideas.

5) Conduct thought experiments to determine the consequences of the potential solutions. What will the future be like if this solution was implemented?

The Research Proposal

6) Create a prioritized short list of potential solutions. Use Tool # 2 *Categorical Selections* or another convergent thinking tool to evaluate the ideas and reduce the number of options to a reasonable set that can be evaluated in more detail. Create MUST HAVE and WANT criteria that can be used to compare the potential solutions. Be optimistic and avoid consideration of constraints or limitations at this stage. Keep a record of the criteria that you used to make the selection and prioritization. You may need to communicate these criteria to stakeholders, or you may wish to revise your criteria in the future. Avoid discarding any ideas or potential solutions. Keep a record of low priority ideas for future reference. Let your ideas incubate for a time. Merge, modify and improve your ideas.

7) Consider using Tool # 3.1 *Expert Panel* or *Tool # 3.2 The Delphi Selection Method* to review the technical details. Ask: Are there existing technologies, processes, services or products that might help to solve the problem? Will new technologies, processes, services or products be available in the future? What new technologies, processes, services or products can we create that will solve the problem?

8) Consider using Tool # 3.3 *Deliberative Forum* to identify areas of tension and areas of support. Gain different perspectives on the topic and obtain feedback from others.

9) Consider the benefits of each potential solution in your analysis: Who will consider the solution to be beneficial? Under what circumstances? Who will consider the solution to be harmful? Under what circumstances? Change is disruptive and this will impact some people negatively. How can the benefits be enhanced? How can the detriments be minimized? What quantitative improvements might be made if the solution was implemented? For example, financial returns or improved quality of life. Do you wish to create financial value? How will this solution create value? How can the value be captured?

10) Conduct a Force Field analysis on your vision (Tool # 9.2 provides details of the process). Consider the factors likely to help vs hinder support for your vision. List the individuals and groups positively and negatively impacted by implementation of your vision.

11) Consider the risk involved with each potential solution in your analysis: Adaptive innovation is relatively low risk. Radical innovation is relatively high risk.

12) Based on your analysis, select one or more potential solutions to propose further.

13) Write a Vision Description that includes the technical details and your analysis of the potential solutions. This summary may be used in grant applications or business proposals.
   - What is the ideal solution to the problem?
   - Who will use or implement this solution? This is a potential customer.
   - What might the future be like if the problem is solved? This provides an estimate of the project's value.
   - Avoid stating HOW this vision might be achieved. Details will be added later.
   - Include a layman's summary: Where are we? Where do we want to go? Why do we want to go there?

14) Write a one sentence Vision Statement: If [CUSTOMER] is able to [DO WHAT] then [SOLUTION]. The Vision Statement is intended to motivate others to support your concept. Make the wording inspirational, but realistic.

15) Next step: Use the Vision Description to identify the scientific phenomena that should be analyzed in detail to determine how the vision might be implemented.

TOOL # 7 THE SCIENTIFIC MODEL

WHAT:

A logical understanding of a scientific phenomenon.

A diagrammatic, iconic or pictorial representation of relationships among empirical observations of a scientific phenomenon.

Why:

To communicate our current scientific knowledge about relationships related to a scientific phenomenon.

To identify gaps in knowledge, missing information and assumptions about relationships.

To predict events and outcomes that might be tested in an experiment or used to implement a solution.

Input:

The Problem Statement and Problem Description from Tool # 5: [WHO] needs [WHAT] because [WHY].

The Vision Statement and Vision Description from Tool # 6: If [CUSTOMER] is able to [DO WHAT] then [SOLUTION].

Output:

Rival scientific models that explain empirical observations made on a scientific phenomenon.

How:

1) Conduct a detailed review of the scientific literature based on your description of the problem and your vision. Critically review the published empirical data and observations that support the current scientific model. Critically review the techniques and methods used to conduct experiments. Check the validity of the statistical analysis of the data. Check for bias in the interpretation of empirical data. Conduct a meta-analysis, if possible. For most scientific topics, the organization of the data is insufficient to conduct a meta-analysis. Assess the importance and significance of the published observations and data.

2) To summarize and visualize relationships, use one or more of the following methods:
    - Use Tool # 7.1 *Is/Is-not* to organize the observed effects of a single cause. You may use the attributes from Tool # 5 *The Problem Statement* as a starting point.

- Use Tool # 7.2 *Four Windows* to organize the observed effects of two potentially interacting causes.
- Use Tool # 7.3 *Ishikawa Diagram* to visualize a single effect that has multiple potential causes.
- Use Tool # 7.4 *Interrelationship Diagram* to visualize complex topics that have multiple distinct effects and multiple potential causes.

3) Alternatively, prepare a table. Label the first columns as Effect (dependent variable), the second as Cause (independent variable) and the third to fifth as Experiment1, Experiment2 and Experiment3, where 1, 2 and 3 correspond to the required criterion to establish the relationship:

1. [CAUSE] is always present when [EFFECT] is observed; there is a correlation between the two variables.

2. The [EFFECT] is observed only when [CAUSE] is present; when [CAUSE] is applied [EFFECT] subsequently occurs.

3. The [CAUSE] is sufficient for [EFFECT] to occur; no other factor is necessary.

4) Complete the table by classifying each putative cause-effect relationship and enter a color or letter code corresponding to the level of support for each of the three criteria in the reported experiments:

A - Proven - criterion met = green.

B - Tentative - some experimental support = yellow.

C - Assumed - not tested experimentally = black.

D - Controversial - contradictory observations = red.

Effect	Cause	Experiment1	Experiment2	Experiment3
Y	A	A	B	D
Y	B	A	D	C
Y	F	A	A	C
Y	G	A	A	A

5) Understand the contradictions and controversial relationships (type D above).

6) Understand the assumed relationships (type C above). What is the

probability that these assumptions are valid? To test for reverse causation, state the opposite of the accepted assumption and speculate on what happens if this reversed assumption is correct. Identify any assumptions that may be challenged.

7) Understand the tentative relationships (all type B above).

8) Use abductive logic and your understanding of cause-effect relationships to prepare a scientific model. Summarize your scientific model in a diagram, flow chart or a story board. This will improve clarity and provide focus. Most people understand concepts better when presented with a visual representation. You model should meet the following criteria:

✓ Make useful predictions.

✓ Describe simple, measurable relationships.

✓ Incorporate current paradigms.

✓ Explain all of the empirical observations and data.

✓ Be dynamic and amenable to change.

✓ Withstand criticism.

9) Include a clear indication of the knowledge gaps and assumptions in your model.

10) Change the assumptions to create rival scientific models that make different predictions. Each rival scientific model must meet the above criteria. State all assumptions in an explicit manner. Propose several new testable hypotheses.

11) If necessary, revise your Problem Statement from Tool # 5 and/or your Vision Statement from Tool # 6 based on your more detailed analysis of the literature and your rival scientific models.

12) Communicate your scientific models to experts and peers. Seek their feedback. On technical topics, seek the advice of experts using Tool # 3.1 *Expert Panel* or Tool # 3.2 *The Delphi Method*. Gain different perspectives on the topic from non-experts using Tool # 3.3 *Deliberative Forum*.

13) Revise the model based on new empirical information as it is published. This is a "living" model that needs to be curated to improve the accuracy and utility of its predictions.

TOOL # 7.1 IS/IS-NOT

This tool organizes the data and information from a literature review for a relatively straightforward topic.

1) List the perspectives, attributes, characteristics or components of a scientific phenomenon as rows in a table. For example, in the life sciences, you might view a topic from the perspective of ecology, organism, tissue, cell, biochemical, molecular and genetic perspectives. Use any perspective and any categorization of the topic that is appropriate to cluster the observations into themes. Tool # 5 *The Problem Statement* identified the attributes of a problem and these may be an initial way to characterize the effects from a single cause.

2) Enter the data and observations collected from your literature review or from previous observation projects in either the Is or Is-not columns.

Example: the effect of increased carbon dioxide on a maize crop.

Carbon dioxide in the atmosphere is used in photosynthesis by plants to make sugar. C3 plants like wheat have a mechanism of photosynthesis that will fix higher amounts of carbon dioxide as the atmospheric concentration increases. C4 plants like maize have a different mechanism to fix carbon than C3 plants. Consequently, maize will respond differently than wheat to increased levels of carbon dioxide in the atmosphere. The response of maize can be summarized in a table schedule as the one below. A separate table can be constructed for wheat allowing a comparison of the two crop's responses to climate change from different perspectives.

a) In the first row, enter the effects of carbon dioxide on maize from the crop perspective; enter the attributes that drought impacts in the Is column and enter the attributes not impacted in the Is-Not column.

b) Repeat this for the effects on other attributes, such as effect on a whole plant, on a single leaf, on a single cell and other perspectives, row by row.

Effect of Increased Carbon Dioxide on Maize		
	Is	Is-Not
Crop		
Whole Plant		
Leaf		
Root		
Cell		
Biochemical		
Molecular		

Notes:
- In this example, it is necessary to define "increased". An option is to specify the amount of increase and to use multiple tables for different ranges to simplify and clarify the summary.

TOOL # 7.2 FOUR WINDOWS

This technique summarizes relatively simple interactions discovered from your literature review among two independent variables that may causes one or more observed effects.

1) Prepare a 2 x 2 table such as the one shown below.
2) From the literature review, identify two putative causes of the observed effects. These may be treatments that have been imposed on a sample or variables that have been corrected with the observed effects
3) Label the columns of the table with one variable (cause) and the rows with the other.
4) Assign a categorical label to each row and column such as Yes versus no; high versus low; up vs down. Alternatively, you can use more than two levels of each variable if you can make an accurate categorical distinction, but do not make the table too complicated.
5) Enter the data and observations (effects) collected from your literature review or from previous observation projects into the four windows.
6) To keep the organization relatively simple in the case of complex phenomena, construct different tables to represent different

perspectives or different types of type of observations, for example ecological, organism, tissue, cell, biochemical and molecular observations.

Example: the effect of drought on a maize crop.

The observations describing the response of a maize crop to drought under high and low temperatures and high and low carbon dioxide levels can be summarized in a Four Windows table in order to make a scientific model that predicts the effect of climate change on a maize crop. Separate tables can be constructed for response at the whole plant, leaf, root, cell and biochemical perspectives.

Effect of Drought on Maize		Temperature	
		High	Low
Carbon Dioxide	High		
	Low		

TOOL # 7.3 ISHIKAWA DIAGRAM

An Ishikawa diagram or fishbone diagram is useful for visualizing multiple causes of a single effect.

1) Draw a fish skeleton as indicated in the diagram below.
2) Write the effect or a problem as the head of the fish.
3) Write the putative major causes of the effect as headings for each of the major fishbones (branches off the central backbone). Use as many fishbones as needed. For a generic problem, the potential causes might be classified and grouped into themes such as people, materials, methods, measurements, equipment or environment. Alternatively, branches may represent scientific perspectives, such as ecological, organism, cell, physiological, biochemical, molecular and genetic topics.
4) Enter the putative minor causes as braces to the major fishbone. Include data and observations from the literature review into the appropriate fishbone categories.

The Research Proposal

5) Reorganize the major causes, minor causes and observations, as needed, or add more major categories.
6) Apply the repeated questions technique to better understand causes and effects. Ask either "Why?" or "How?" to each of the relationships to dig deeper and understand the root cause of the effects.
7) Ask "Why Else?" or "How Else?" to expand your perspective and think laterally.
8) Ask "So What?" to understand the relative significance of the different observations. This relatively modest thought process allows you to critically analyze the observations and develop a better understanding of the phenomenon, especially the putative cause-effect relationship.
9) Determine (usually a subjective estimate or guess) the relative impact of each cause on the effect. Group related observations into a hierarchy using an affinity diagram, tree diagram or dendrogram.
10) Optional: Use *Tool # 2.3 Pareto Analysis* to select the putative causes that contribute 80% of the effect.

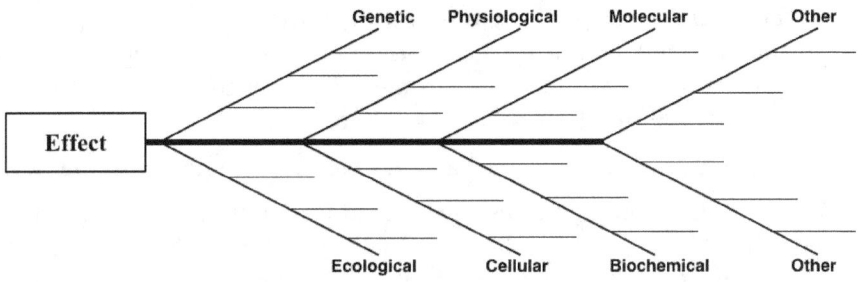

You may also reverse the fishbone diagram to show multiple effects from a single cause. For example, the effects of increased carbon dioxide levels in the atmosphere might be shown on different ecosystem.

TOOL # 7.4 INTERRELATIONSHIP DIAGRAM

An interrelationship diagram is useful for visualizing complex phenomena, in which multiple effects are attributed to multiple causes. An affinity diagram is a variation commonly used in business and project management to show relationships.

1) Prepare a table with 5 columns. See the example below.

2) In column 1 write a list of the putative major causes of the observed effects. These are called independent variables because they do not depend on other variables. These are the classes of treatments that are imposed on a sample, or the classes of measured observations that have a putative relationship with another variable (e.g. biochemical pathway or regulatory network).

3) In column 5 write a list of the observed effects. These are called dependent variables because they depend on or are caused by the independent variables. These are measurements or observations made following treatment.

4) Optional: In column 2 write the minor causes, as needed to create clarity. These are components or attributes of the major cause. These are the quantitive variables

5) Optional: In column 4 write the minor effects or the precise measurements used to collect data or observations, as needed to create clarity..

6) In column 3 draw an arrow to indicate a relationship between the independent and dependent variables; this is a putative cause-effect relationship.Variables can be linked by the number or strength of their relationships. The relationships may be based on subjective analysis, but quantitative criteria, such as correlation, is preferred. Your confidence in the experimental evidence can be indicated, by either the strength or color of the line.

7) Group ideas and putative cause-effect relationships into themes.

8) Apply the repeated questions technique to better understand causes and effects. Ask either "Why?" or "How?" to each of the relationships to dig deeper and understand the root cause of the effects.

9) Ask "Why Else?" or "How Else?" to expand your perspective and think laterally.

10) Ask "So What?" to understand the relative significance of the different observations. This relatively modest thought process allows you to critically analyze the observations and develop a better understanding of the phenomenon, especially the putative cause-effect relationships.

11) Determine (usually a subjective estimate or guess) the relative impact of each cause on each effect. Group related observations into a

hierarchy using an affinity diagram, tree diagram or dendrogram.

12) Optional: Use *Tool # 2.3 Pareto Analysis* to select the putative causes that contribute 80% of each effect.

Independent Variables			Dependent Variables	
Subject	A			
	B			
	C		X	Measurement
Treatment	D			
	E		Y	Measurement
Enivronment	F			
	G		Z	Measurement
Method	H			
	I			
	J			
Material	K			
	L			

In the above example, the twelve independent variables are different subjects, treatments, environments, methods and materials that affect three dependent variables X, Y and Z which can be measured quantitatively. Some independent variables have no effect, or their effect was not measured. A affected only X, whereas B D F and G affected Y, and G affected both Y and Z.

TOOL # 7.5 THE GAP STATEMENT

1) Identify the gaps in your model that represent unknown, assumed or tentative cause-effect relationships.

2) Identify the assumptions that are critical to your vision but differ among rival scientific models

3) Identify knowledge gaps in the rival scientific models that are significant obstacles to achieve your vision.

4) Write as many Gap Statements as necessary: **If we know [WHAT], [CUSTOMER] will be able to [DO WHAT].**

5) The Gap Statement may become the Problem Statement for an Observation project. Does this gap merit an Observation project to establish relationships? Can the gap be filled with a reasonable

assumption?

TOOL # 7.6 THE PREDICTION STATEMENT

1) Use deductive logic to make predictions based on your rival scientific models.
2) Write a simple Prediction Statement: **Because [A] causes [B], then [X] will occur.**
3) Write a more complex Prediction Statement if the effect is caused by more than one variable: **Because [A] causes [B], and because [C] causes [D], and because……., then [X] will occur.**
4) Add as many premises and cause-effect relationships as necessary to each prediction.
5) Make multiple or alternative Prediction Statements from your rival scientific models.
6) Draw the Prediction in a flow chart or relationship diagram to communicate your concept effectively to sponsors, customers, and your research team.

TOOL # 8 THE OBJECTIVE STATEMENT

WHAT:

The goals of a research program, research project, experiment or task.

A statement of what knowledge will be learned to achieve the vision.

Why:

To establish and communicate the knowledge to be created by the research project.

To define the success criteria of a research project.

To recruit and to motivate a research team and stakeholders.

To empower a research team to make decisions.

Input:

The Problem Statement and Problem Description from Tool # 5: [WHO] needs [WHAT] and [WHY].

The Vision Statement and Vision Description from Tool # 6: If [CUSTOMER] is able to [DO WHAT] then [SOLUTION].

Rival Scientific Models from Tool # 7.

Gap Statements from Tool # 7.5: If we know [WHAT], [CUSTOMER] will be able to [DO WHAT].

Prediction Statements from Tool # 7.6: Because [A] causes [B], then [X] will occur.

Output:

Outcome Description to be used in business proposals and grant applications.

Objective Statement for each research program, research project, experiment or task: **Our team will learn [WHAT]**.

Objective Statement for an experiment or a task within a research project: **[WHO] will accomplish [WHAT] by [WHEN]**.

How:

1) Using Tool # 1 *Brainstorming*, search for ideas that fill the gaps in your scientific model. Collect as many ideas as possible.

2) Use convergent thinking techniques to evaluate the ideas and reduce the options. Prioritize and select the "best" options using Tool # 2 *Categorical Selections* or another convergent thinking tool. Consider using Tool # 3 *Review Panel* to gain the perspective of others. Use

Tool # 4 Group Voting to make a final selection.

3) List the knowledge gaps based on their importance to your vision in a logical sequence. Each gap becomes a potential research project.

4) Prepare an Outcome Description: What do you wish to achieve? What knowledge will be created? Who needs this knowledge, data or information? What will change if this knowledge is acquired? Is the team identified? Is there a deadline?

5) Simplify your Outcome Description to a one sentence Objective Statement: **Our team will learn [WHAT]**. The statement should meet these criteria:

   ✓ Succinct.

   ✓ Simple and easily understood.

   ✓ Challenging but possible.

   ✓ Defined by success criteria that can be measured transparently.

   ✓ Time-specific (optional).

6) Write an Objective Description for each knowledge gap in your scientific model.

7) Determine the consequences if the objective is achieved. Compare the desired outcomes relative to the undesired outcomes. Who will consider this to be beneficial? Under what circumstances? These individuals or groups are potential supporters or customers. Who will consider this to be harmful? Under what circumstances? Those individuals or groups who may be negatively impacted will resist or directly oppose your research project. Pay special attention to these potential opponents.

8) Revise the Objective Description to maximize the desired outcomes and to minimize the undesired outcomes or potentially harmful effects. This will increase its impact and importance and reduce potential criticism by opponents.

Notes:

- The Objective Statement may be created by an individual but team-based approaches usually create more options and better ideas.

- The Objective Statement simply indicates WHAT will be learned, not HOW.

- An Objective Statement can be written for a research program, research projects, individual experiments or even tasks. The difference is one of scope, not process.

- Before funding is approved, the Objective Statement is malleable and can be revised iteratively as new information is gathered, new perspectives considered, and sponsors contacted.
- Once the project is funded, the Objective Statement is fixed! The strategy and tactics to achieve the objective may change, but not the objective. Achieving the objective marks the successful endpoint of a research project. Do not yield to the temptation to revise your objective after you complete the experiments. This simply leads in the long term to mediocrity because you will fail to learn, to improve and to excel.

TOOL # 9 THE RESEARCH CONCEPT

WHAT:

A succinct summary of a potential research project.

Description of the knowledge required to achieve an objective and implement a vision.

Why:

To identify the essential accomplishments of a research project prior to budget approval.

To communicate a concept that gains the support of stakeholders, customers and sponsors.

To enable critical evaluation and selection from a pool of alternative concepts for research projects.

To respond quickly when funding or business opportunities arise.

To guide and to empower a research team to make strategic and tactical plans.

Input:

The Problem Statement and Problem Description from Tool # 5: [WHO] needs [WHAT] because [WHY].

The Vision Statement and Vision Description from Tool # 6: If [CUSTOMER] is able to [DO WHAT] then [SOLUTION].

Rival Scientific Models from Tool # 7.

Gap Statements from Tool # 7.1: If we know [WHAT], we will be able to [DO WHAT].

Prediction Statements from Tool # 7.2: Because [A] causes [B], then [X] will occur.

Objective Statements from Tool # 8: We will learn [WHAT].

Output:

A document that can be submitted to decision-makers, sponsors and customers.

A succinct summary that can be used to recruit support for a research project.

How:

1) Prepare a draft Research Concept document by merging the Problem Description (Tool # 5), Vision Description (Tool # 6), rival Scientific

Models (Tool # 7), one or more Gap Statements (Tool # 7.5), one or more Prediction Statement (Tool # 7.6) and one or more Objective Statements (Tool # 8). Indicate why this problem merits attention and the potential benefits. Indicate how your vision will solve the problem. Indicate what you need to learn.

2) Use Tool # 1 *Brainstorming* to identify the accomplishments required to achieve the proposed objectives from Tool # 8. Indicate what the project needs to accomplish to create the required knowledge. Avoid listing activities and tasks; focus on accomplishments. In other words, focus on what you want to achieve, not what you need to do. Avoid making tactical plans for experiments.

3) Review the required accomplishments in a group meeting with the research team members, including potential collaborators and stakeholders (if known), to gain their support and commitment. Ask: will these accomplishments create the knowledge needed to achieve the objective? Revise the Research Concept as needed.

4) Select the three most critical accomplishments that you must achieve using Tool # 2 *Categorical Selections* or *Tool # 4 Group Voting*. What accomplishments must be done to achieve the objective? In project management terminology, what is In-Scope? Other accomplishments are considered to be nice-to-have. These accomplishments will create value, reduce risk or provide an opportunity, but are not critical to success.

5) Determine what the potential research project will not do. What will the research project not accomplish? In project management terminology, what is Out-of-Scope? Is support needed from service providers or external collaborators to achieve critical accomplishments? How can this support be obtained? Revise the Research Concept as needed.

6) Develop a preliminary budget based on your experience and a guess at the required resources. This will be revised later after tactical planning is completed.

7) Identify potential customers for your Research Concept. Use *Tool #11Customer Assessment* as a guide.

8) Identify potential sponsors for your Research Concept. Your R&D organization will be the best source of information on how to select and apply to local granting agencies and other sponsors.

9) Improve the research concept by considering potential customers and sponsors. Use one or more of these tools: Tool # 9.1 *SWOT*

Analysis, Tool # 9.2 *Force Filed Analysis and* Tool # 9.3 *Win Conditions.* Revise the Research Concept as needed.

10) Use Tool # 9. 4 *The Research Project Skinny* to draft a succinct statement describing the research project that may be used in communication to decision-makers.

11) Add the Research Concept to a portfolio of potential research projects. Every scientist and every R&D organization has a pool of potential research projects. Let the Research Concept incubate in your portfolio. This is a dynamic document; repeat the above process as needed when new scientific knowledge, new technology or new opportunities arise.

TOOL # 9.1 SWOT ANALYSIS

1) Conduct a SWOT analysis on a Research Concept document in your portfolio considering various perspectives. Use Tool # 1 *Brainstorming* to identify the relevant attributes in the following categories. Add other categories that are relevant to the research project or scientific discipline.

Research team and R&D organization - Include skill of research staff, experience in conducting this type of experimentation, facilities, equipment and any competitive advantage relative to other R&D organizations.

Scientific - Include the importance and impact of the problem, uniqueness of the scientific model, confidence in the assumptions in the scientific model, and intellectual property that gives "freedom-to-operate".

Sponsor - Include their level of support, financial stability and commitment.

Customers - Include their ability to use the knowledge created, enthusiasm, need and ability to use the knowledge.

Marketplace - If there is a financial objective, include the possibility of market changes.

Risk - Include the possibility that a critical experiment or task will fail or be delayed.

2) For each perspective, identify:

Strengths – the positive and helpful advantages in the Research Concept.

Weaknesses - the negative and harmful disadvantages in the Research Concept. Include potential risk events that may occur and their impact.

Opportunities – events that may occur in the external environment that will add value. Include the opportunity to create intellectual property, publish in high quality journals or have a greater than expected financial return.

Threats – events in the external environment, such as the actions of competitors, that may impede success. Include the possibility that competitors will publish or patent similar results, develop blocking intellectual property, or receive competitive grant funding.

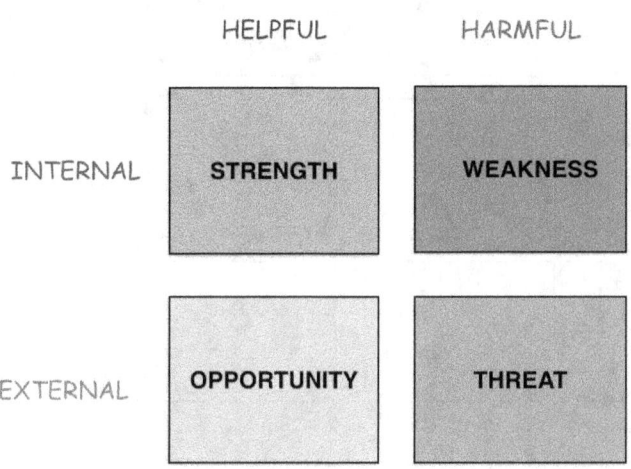

3) Compare the helpful versus harmful SWOT attributes and understand the factors influencing each attribute. Strengths that support a research project and Opportunities in the external environment are considered helpful. Weaknesses and Threats from the external environment that impede success are considered harmful. List the attributes as bullet points in a diagram, such as the one above, or list in a table. Ask "Why?" or "How?" for each

attribute to understand the root cause of the assessment. In a group meeting, rank the bullet points from most to least important using Tool # 4.1 *Multivoting*. Use Tool # 2.3 *Pareto Analysis* to focus on the attributes that will have the greatest impact. Discard attributes that have a relatively minor impact.

4) Use Tool # 1 *Brainstorming* to search for ideas that may capitalize on strengths and opportunities to minimize weaknesses and threats. Here are some suggestions for questions to ask the group:
   - How can strengths be used to capture opportunities?
   - How can strengths be used to minimize the possibility of a threat?
   - How can opportunities be used to minimize weaknesses?
   - How can weaknesses be minimized to avoid threats?

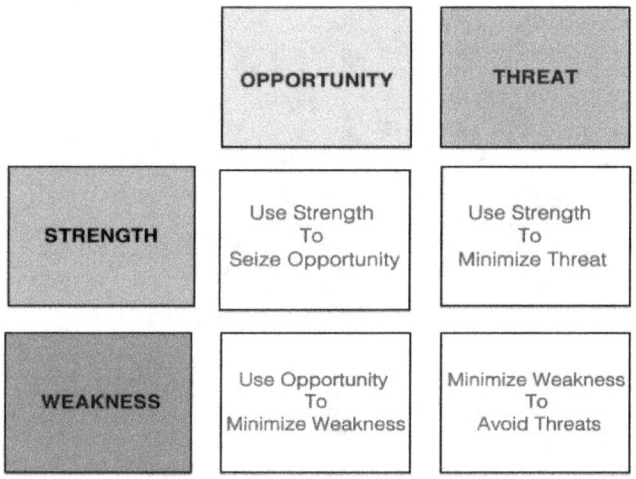

5) Revise your Research Concept to incorporate these ideas, as appropriate.

6) Use the SWOT analysis to compare several strategies, approaches or Research Concepts. Identify those with the most helpful and the fewest harmful attributes. Use the identified attributes in Tool # 2 *Categorical Selections* to make a selection.

TOOL # 9.2 FORCE FIELD ANALYSIS

1) List the factors likely to influence a decision-maker's support or opposition to the Research Concept. There may be several specific concerns in each of these categories. Here are some examples to consider but add other relevant factors.

- Ethical and moral issues.
- Legal and regulatory issues.
- Budget and resource issues.
- Staffing and expertise issues.
- The importance of the problem.
- Acceptance of your vision.
- Tolerance of the change or innovation described in your vision.
- The quality of the information used to construct a scientific model.
- The interpretation of specific facts, data and observations.
- The logic used to construct the rival scientific models.
- The validity and acceptance of the scientific assumptions in a model.
- The logic used to propose hypotheses.
- The prioritization of hypotheses to be tested experimentally.
- Support for the current paradigms and pet theories.
- An individual's job security, status or organizational stability.
- Self-interest based on how the project might impact an individual's own interests.

2) Prepare a table, listing the influencing factors in the first column.

3) List the decision-makers who will decide whether to support or oppose your research project as the headings of each column in the table.

4) In each cell of the table, indicate whether this decision-maker will support (plus), oppose (minus) or be neutral (zero) in regard to this factor. This rating may be based on your informed judgement, an educated guess or preferably from a conversation.

5) Revise the Research Concept to capitalize on positive forces and to minimize negative forces from the perspective of the sponsor.

6) Optional: Sum the ratings to identify the most important factors to adjust in a way that will increase the positive forces and reduce the negative forces.
7) Optional: Select among several Research Concepts for the concept with the largest positive force or greatest potential support of the decision-makers.

TOOL # 9.3 WIN CONDITIONS

1) List DeCarlo's seven project win conditions in the first column of a table:

   Schedule means meeting deadlines and milestone dates.

   Budget means controlling costs to stay within the allocated grant or budget.

   Scope means acquiring the required knowledge.

   Quality means acquiring rigorous knowledge.

   Return on investment means meeting the financial expectations of the sponsor.

   Stakeholder satisfaction means that all of the people involved or impacted by the research project are satisfied.

   Team satisfaction means providing benefit to the research team.

2) Add any other win conditions that may have been identified in Tool # 9.1 *SWOT Analysis*, as needed. For example, it may be essential to mitigate against a potential Threat by applying one of the project's Strengths.
3) Add any personal win conditions of key decision-makers identified in Tool # 9.2 *Force Field Analysis* that may impact their support, as needed.
4) In the second column of the table, add specific details, quantifiable measurements or criteria that describe the win condition in more detail. Add ideas from the research team to ensure that all perspectives are included. An option is to use Tool # 1 *Brainstorming* to search for ideas and details; this may stimulate discussion and awareness within the team. Add details from Tool # 9.1 *SWOT Analysis*, Tool # 9.2 *Force Field Analysis* or feedback from other sources.
5) Determine the relative importance of the win conditions in a meeting

of the research team. Consider different perspectives. Rank the win conditions from most to least important using Tool # 4.1 *Multivoting*. It is important that the research team discuss priorities and understand the consequences of failing to meet any of these win conditions.

6) In the third column of the table, indicate the relative importance of each of these win conditions to the research team.

7) In the fourth column of the table, indicate the relative importance of each of these win conditions to potential sponsors.

8) In the fifth column of the table, indicate the relative importance of each of these win conditions to potential customers. Customers in this case may be the scientific community, in which case the ranking is based on scientific criteria.

9) Revise the Research Concept as needed considering the relative ranking of the win conditions if there is a conflict among alternatives.

10) In the sixth column indicate the overall most important win condition to achieve. This becomes the focus for any further refinement of the Research Concept to ensure that this win condition is achieved. This win condition will be important to monitor as the experiments and research project are conducted.

TOOL # 9.4 THE RESEARCH PROJECT SKINNY

1) Organize a meeting. Invite the research team and key stakeholders of the research project. It is optional to include decision-makers.

2) Clarify the objective of the meeting.

3) Assign a Facilitator to lead the meeting and stimulate the discussion. This is not necessarily the Principal Investigator or Project Manager.

4) Facilitator assigns meeting roles to other participants: Time Keeper watches the time as allocated to each topic and informs the group. Scribe compiles ideas on a white board to facilitate discussion. Note Keeper records the ideas collected from the group. Process Checker ensures that agreed process is followed.

5) Prior to the meeting, the participants receive all relevant background information, including the latest version of the Research Concept from Tool # 9. The participants conduct their own independent

review of the topic.

6) The Facilitator asks a series of questions: WHO will do this work? WHAT must be learned? WHO is the customer that will use the knowledge?

7) Discuss each question separately. There will likely be several answers to each question. Reach a consensus decision (using Tool # 4.2 *Consensus*) to select only one option. If there is more than one alternative answer, consider dividing the Research Concept to create related but independent research projects. If there is overlap or confusion on who is doing the work, what is to be learned or who will use the knowledge, the project will lack focus and clarity. It might be better to split the concept into several research projects.

8) Write the first sentence of the Research Project Skinny: **[WHO] will learn [WHAT] for [WHOM].** Discuss further until there is consensus agreement.

9) The Facilitator asks a second question. WHAT must be accomplished for the customer to consider the research project successful? This might be a long list. Avoid listing all of the accomplishments and requirements, but focus on those that have no further dependencies (i.e. the endpoint of a series of tasks). Prioritize the list using *Tool # 4.1 Multivoting*.

10) Write the second sentence of the Research Project Skinny: **This research project will be considered finished when [WHAT] has been accomplished.** Discuss further until there is consensus agreement.

11) The Facilitator asks a third question: WHAT will this knowledge do? Alternatively, how will the customer use this knowledge? This might be a long list, but prioritize and select the most important one or two applications.

12) Write the third sentence of the Research Project Skinny: **This knowledge will [DO WHAT].** Discuss further until there is consensus agreement.

13) Write the complete Research Project Skinny:

    [WHO] will learn [WHAT] for [WHOM].

    This research project will be considered finished when [WHAT] has been accomplished.

    This knowledge will [DO WHAT].

14) Discuss the final version until there is consensus agreement. It may be necessary to review and revise the first or second sentences to make a consistent statement.

15) Revise the Research Concept document based on the discussion.

TOOL # 10 FINANCIAL ANALYSIS

WHAT:

An analysis of the financial returns from an investment in the creation of knowledge.

Why:

To justify the cost of a research project to potential sponsors.

Input:

The Research Concept from Tool # 9.

The Problem Document from Tool # 5.

Output:

An estimate of the financial cost of a problem.

An estimate of the budget required to conduct a research project.

An estimate of the cost to implement a solution to a problem.

An estimate of the financial returns from a research project that creates knowledge about a problem.

How:

Two estimates are required to calculate the potential financial returns from a research project: the cost of the problem and therefore the potential returns from a solution and the budget for the research project to acquire knowledge that may solve that problem.

1) Estimate the value that is lost due to the problem in dollars annually.

   a) List the people, industries and issues that are affected by the problem.

   b) For each, indicate the type of loss that occurs, for example, loss of production, loss of life, loss of property, environmental degradation, etc.

   c) For each, estimate the value of the loss annually.

   d) Sum the individual losses to estimate the total economic loss annually.

2) Estimate the potential for added value. Note that many commercial R&D organization often conduct an Observation Project to make accurate estimates of value and determine how to capture value from an innovation. This is only a high level overview of the process.

The Research Proposal

   a) Determine the total annual value of the innovation or product being produced.
   b) Determine the value chain from raw material to commercial product sold to the consumer.
   c) Identify the processes or industries that participate in the value chain.
   d) Determine the proportion of value that is added at each step in the value chain using either dollars or percentage of final value.
   e) Identify the process in the value chain where your innovation to solve the problem will add value.
   f) Estimate quantitatively how much value will be added and how that value will be captured.
   g) Determine the added value annually based on current levels of annual production and consumption.
3) Use either the economic loss or the potential value added to estimate the annual cost of the problem.
4) Develop a preliminary Budget for a Research Project. Note this is only a guess at this time based on your experience. This preliminary budget estimate will be revised later after detailed tactical planning is completed.
   a) Estimate the duration of a research project..
   b) Estimate total annual personnel costs.
   c) Estimate the number of scientists who will be needed to complete the project within the specified duration.
   d) Determine the budget cost for each on the scientists considering salary, benefits and overhead costs.
   e) Determine major equipment costs. Consider the infrastructure resources and services required. Determine the capital costs, rental fees, service fees and operating costs for each.
   f) Estimate the annual operating costs for materials and supplies, travel, publication fees, legal fees.
   g) Total all operating costs including personnel, equipment and supplies.
   h) Add the overhead costs from your R&D organization for administration, resources, utilities and other support.
   i) Add a contingency budget based on your confidence in the

preliminary estimate. I suggest to use at least 10% contingency if you have high confidence in your estimate and use a higher rate if you have less confidence. Avoid underestimating research project costs at this preliminary stage.

5) For an Observation, Modeling or Discovery project, that will not produce a tangible prototype or product, use Tool # 10.1 *Return on Investment*.

6) For a Development project that will produce a tangible prototype or product, use Tool # 10.2 *Net Present Value*.

TOOL # 10.1 RETURN ON INVESTMENT

1) Use either the annual economic loss or the annual potential value added as an estimate of the maximal annual potential returns. Use the preliminary budget as an estimate of the investment.

2) Create a table. Label the first column as year.

3) Estimate the following dates in a timeline and enter into the first column of a table.

   a) Start date of the research project.

   b) End date of research project.

   c) Date of the first implementation or sale (e.g. first economic return).

   d) Date of full implementation or maximal sales. Assume that there will be a period of introduction, e.g. a lag period before acceptance.

   e) Last date of full implementation. Assume that there will be a period of full implementation, e.g. a period of full acceptance and maximal market sales, followed by a decline as other products, process or innovations replace this one.

   f) Date that your proposed innovation will cease to have any added value.

3) Enter the anticipated expenses or income for each year in the second and third columns.

4) Estimate the Total Value by summing the returns for the duration of the innovation from start to end.

5) Estimate the Total Research Cost from start to end.

6) Optional: add the costs of implementation of an innovation, such as infrastructure, materials and training, to estimate Total Cost. Sometimes research costs are relatively small compared to scale-up and implementation costs. Enter Production Costs in the spreadsheet as a separate column.
7) Calculate Net Return = Total Value – Total Cost.
8) Calculate Return on Investment = Net Return / Cost as a percentage

Example Table:

Year	Research Cost	Production Cost (optional)	Revenue
1			
2			
3			
n			
Total			
ROI	(Total Revenue – Total Research Cost)/ Total Research Cost expressed as a percentage		

Notes:
- See text for a sample calculation.
- ROI can be used for Observation, Modeling or Discovery projects but the duration from research to implementation is too long to make anything but guesses about revenue.
- Use ROI to compare the relative economic value of different research projects.
- ROI is best used in hindsight to estimate the value of past research projects because the costs and revenues can be more accurately estimated.
- See Wikipedia and Investopedia for more detailed instructions on the calculation of ROI.

TOOL # 10.2 NET PRESENT VALUE

1) Use the table from Tool # 10.1 *Return on Investment* that lists

anticipated costs and revenue.

2) Add two columns to the table: Annual Cash Flow (R_t) and Discounted Cash Flow.

3) From the start of the research project to the last year of returns, calculate Cash Flow (R_t) for each year (t) = Revenue (t) - Cost (t). For the first years during the research project, Cash Flow (R_t) will be negative and in later years after research is completed and products are sold, Cash Flow (R_t) will be positive.

4) Estimate the Discount Rate (i). This is the expected annual rate of return on an investment with similar risk. The expected rate of return on bonds or stock dividends is currently less that 4%, but research has much higher risk. As a preliminary estimate, you can use 10% but adjust based on the estimated risk in your research project and current investor expectations.

5) Calculate the Discounted Cash Flow for each year (t) = $R_t / (1+i)^t$. Enter the calculation for annual Discounted Cash Flow in the fifth column of the spreadsheet.

6) Calculate Net Present Value as the sum of all the annual Discounted Cash Flows.

Example Table:

Year	Research Cost	Production Cost	Revenue	Annual Cash Flow	Discounted Cash Flow
1					
2					
3					
n					
Total					
NPV	Total Discounted Cash Flow as $				

Notes:
- See text for a sample calculation.

- NPV is an inaccurate estimate of the value of Observation, Modeling or Discovery projects because the duration from research to implementation is too long to make guesses about revenue. Use ROI in Tool # 10.1 instead to compare the relative economic value of different research projects.
- For a Development project that is in the early stages of implementing a solution to a problem, NPV is only a wild guess on its potential value. As such, it should only be used to guess at the relative value of different Development projects with similar risk of success.
- See Wikipedia and Investopedia for more detailed instructions on the calculation of NPV.
- The Discount Rate in the NPV calculation can be based on the risk tolerance of the sponsor. Successfully introducing a product into the marketplace depends on many factors beyond acquiring knowledge. See Investopedia for some of the concerns about Discount Rate.

TOOL # 11 CUSTOMER ASSESSMENT

WHAT:

An analysis of the expectations of the individuals and groups who might use the knowledge created.

The first step towards defining a strategy and a tactical plan for your research project.

Why:

To understand WHO wants WHAT by WHEN.

To establish the deliverables for a research project.

Input:

The Research Concept from Tool # 9.

Output:

A list of all customers of a research project.

A list of deliverables from a research project for each customer.

A list of undesirable outcomes that must be avoided by the research project.

How:

1) Create a Customer Assessment Table using the Research Concept from Tool # 9 as a guide. Potential customers will use the knowledge created by your research project

2) In the first column of the table, list the category of potential customers. Be inclusive at this stage; you can remove customers later if the initial assessment was in error. Use the list of customer categories given as a guide. Add other categories that are specific for your research project or your R&D organization. It is unlikely that a research project will have customers in all categories but will likely have more than one.

3) Identify individuals whenever possible in the second column. There may be more than one individual customer in each category.

4) Decide who is the most important or primary customer, i.e. the primary customer from Tool # 9.4 *The Research Project Skinny*. Review this selection decision and revise if needed by consensus (using Tool # 4.2 *Consensus*). Assign the primary customer an importance of high in the third column.

5) Decide who are the secondary and who are the minor customers.

The Research Proposal

Assign the remaining customers a relative importance rating of medium or low.

6) Discuss the Research Concept with the primary customer. Ask the following questions:
   - What knowledge do you expect to be produced by this research project?
   - Are there other outcomes (deliverables) that you might want from this research project?
   - When do you need delivery?
   - What outcomes do you want to be avoided? What outcome would be harmful?
   - What will make this research project successful in your opinion?
   - How can this research project be improved to make it more successful?

7) From your conversation, confirm that this is your primary customer.

8) Ask similar questions to each of your potential secondary customers. From these conversations, complete the table to indicate the knowledge or deliverable (product from the research) that each customer wants, when they want it and how they want it delivered. In the last column, indicate what each of the customers wants to avoid.

		Deliverables			
Category	Individual	Importance (H M or L)	Deliverable	Date Expected	Avoid

9) Identify any potential conflicts. These will occur when an outcome is desired by one customer but considered to be harmful by another customer.

10) Determine how to mitigate any potential conflict to meet the expectations of both customers. Can this project meet the needs of both customers?

11) If conflicts cannot be resolved, split the Research Concept into two or more research projects that have interdependent activities but separate deliverables. OR remove any customer whose needs and expectations conflict with those of the primary customer.
12) Revise the Research Concept in Tool # 9 based on your customer assessment as needed.

EXAMPLE CUSTOMERS

Category	Description
Scientific Community	The community includes scientific journals, scientific societies, knowledge databases and other archives of scientific knowledge that are available to everyone.
Decision-Maker	This customer uses the knowledge from your research project to make a decision. This may be a Go/No-go decision to proceed with the next phase of research, product development or marketing that involves investment or allocation of resources.
Teacher	This customer trains others in a scientific discipline, a business enterprise or an industry. For example, professors, science writers, technical consultants, educators and professional trainers teach others the current state-of-the-art and technical skills.
Student	This customer is acquiring skills and knowledge to advance their professional careers. This group includes graduate students who are learning how to do research in their scientific discipline, established professionals who seek promotions and career advancement and researchers who are driven by curiosity to seek knowledge and to advance their understanding of science.
Technology Transfer	This office in the university or government agency administers the invention policy of your R&D organization. In commercial organizations this is the patent attorney who files your applications for intellectual property.
Salesman	This customer uses information and knowledge to promote the sales of a product though the marketing department of a business.
Regulator	This customer uses your information and knowledge to create or to confirm legal/regulatory requirements. For example, the Regulator may require information on efficacy and safety of a new product. The Regulator may use knowledge to establish regulatory requirements. New knowledge may require a change to those regulatory requirements.
Businessman	This customer is operating an existing enterprise and incorporates your knowledge into a business to improve its performance or to reduce costs
Fixer	This customer is seeking to overcome a major hurdle or to repair a broken part in an existing business or product. The Fixer uses your knowledge to turn a nonfunctional process into a viable business or product.

Entrepreneur	The Builder takes your knowledge and/or inventions and turns them into practical applications and products. An entrepreneur forms startup companies and creates disruptive technologies. An entrepreneur creates new processes, new businesses or new products that replace the existing ones. The Builder conducts a Development project to develop an innovative product from your discovery.
Contractor	These very special customers also sponsor your research. A legal contract between the contractor and your R&D organizations specifies details of the project and the required deliverables. A fee-for-service agreement specifies the deliverable that is communicated in the form of a confidential written report.
Investor	These are special customers who as sponsors expect a financial return on their investment in research from your intellectual property, inventions, functional services (e.g. diagnostics) or tangible assets. Confidentiality and protection of intellectual property is critical to these customers.

TOOL # 12 SELECTING RESEARCH PROJECTS

WHAT:

A process that analyzes a set of Research Concepts to select one to submit to a sponsor for funding.

Why:

Several strategies or approaches to solve a problem that are described in Research Concepts have different potential to be funded by a sponsor.

Input:

Several Research Concepts from Tool # 9.

The selection criteria to be used by the sponsor.

Output:

A decision on which Research Concept to propose to a specific sponsor for funding.

How:

1) Use Tool # 12.1 *Understanding Sponsors* to make the selection if the potential sponsor has clearly defined selection criteria that must be addressed. This is more common in a academic R&D organization for applications to granting agencies.

2) Use Tool # 12.2 *Success Zones* to make the selection if the potential sponsor wishes to capture opportunities without predefined selection criteria. This is more common in a commercial R&D organization for applications to capture business opportunities.

TOOL # 12.1 SPONSOR ASSESSMENT

1) List the selection criteria that will be used by the sponsor. These may be educational, technical, scientific and/or business criteria. Your R&D organization is the best source for this sponsor-specific information. Divide the criteria into MUST HAVE (categorical absolute rating of yes or no) and WANT (relative comparative rating) categories.

2) Rate each Research Concept from Tool # 9 for the MUST HAVE criteria of the sponsor. This categorical rating eliminates any concept that fails to meet all of the criteria.

3) Rate each Research Concept from Tool # 9 for the WANT criteria of

the sponsor. This relative rating will create a ranking. Scientific merit is often a WANT criterion. These criteria will be assessed by a selection committee based on their subjective evaluation, which may be different than yours. Attempt to view the proposal from their perspective to make these ratings. This ranking is ideally done by a team using Tool # 4 *Group Voting*, but by necessity may be used by an individual.

4) Using Tool # 9.1 *SWOT Analysis* and Tool # 9.2 *Force Field Analysis* list the Limitations and Constraints for each Research Concept.

5) If the selection criteria of the sponsor include any financial impact considerations, use Tool # 10 *Financial Analysis* to estimate Return on Investment or Net Present Value.

Notes:
- A grant application to an external funding agency will require a strategic and tactical plan for your research proposal. The use of research project management practices in the detailed planning of experiments is described in the next volume of the Research Project Management series *The Research Plan*.

TOOL # 12.2 SUCCESS ZONES

1) This comparison may be used with Research Concepts from Tool # 9 to compare concepts with different visions, scientific models or objectives, but works equally well during the planning stage to select the best strategies or tactics to use to test hypotheses.

2) Using Tool # 9.1 *SWOT Analysis* and Tool # 9.2 *Force Field Analysis* list the Limitations and Constraints for each Research Concept.

3) Rate each concept for probability of success as relatively high, medium or low based on your assessment of Limitations and Constraints. A set of questions can distinguish among the concepts for their probabilities of scientific success. Here are some examples:
    - Are the assumptions in the scientific model valid?
    - Are there significant risks?
    - Will the experiments create knowledge regardless of the results?
    - Is the budget sufficient?

4) Using the information from Tool # 7 *Scientific Model*, Tool # 7.5 *The Gap Statement* and Tool # 7.6 *The Prediction Statement*, rate each concept for scientific impact as relatively high, medium or low. A set

of questions can distinguish among the concepts for importance or impact. Here are some examples:
- Might this knowledge be used to solve the problem?
- Might the results be published in a high impact journal?
- Might this knowledge create a new scientific paradigm?
- Might this research create intellectual property?

5) If the selection criteria of the sponsor include any financial impact considerations, use Tool # 10 *Financial Analysis* to estimate Return on Investment or Net Present Value.

6) The ratings and selection may differ among individual sponsors and customers. Consider repeating the evaluation using a different perspective or different set of guidelines representing different individuals on the selection committee of the sponsor.

7) Select the concept with highest probability of success and highest impact to propose to the sponsor.

CITATIONS AND NOTES

Introduction

1. The work breakdown structure is described in any project management manual in the Further Reading section. Two good descriptions are given in: Brown, K., & Hyer, N. (2009). Managing Projects: A Team-Based Approach. McGraw-Hill/Irwin; and in Lewis, J. (2010). Project Planning, Scheduling, and Control: The Ultimate Hands-On Guide to Bringing Projects in On Time and On Budget, Fifth Edition. McGraw-Hill.
2. The scientific method is much more than this simple work breakdown structure. The scientific method is a philosophy not a workplan, but the sequential tasks are similar. See Further Reading for references on the philosophy of the scientific method.
3. Several of the research project management tools are available online at https://www.scientiststoolbox.com/.

Expectations of Research Projects

1. University of California, Berkley hosts a website "Understanding Science 101" at https://undsci.berkeley.edu/article/0_0_0/us101contents_01
2. NSF reports on the public's attitudes and understanding of science at https://nsf.gov/statistics/2018/nsb20181/report/sections/science-and-technology-public-attitudes-and-understanding/introduction. For another survey of people's attitudes towards science see the report from Science Counts at https://www.sciencecounts.org/wp-content/uploads/2019/02/ReportBenchmark.pdf. Other surveys are available at https://sciencecounts.org/research/. Welcome reports on "The world's largest study into how people around the world think and feel about science and major health challenges" and other surveys are available at https://wellcome.ac.uk/what-we-do/reports?&field_topic[17]=17.
3. Berkun, S. (2010). The Myths of Innovation. O'Reilly Media. Burkus, D. (2013). The Myths of Creativity: The Truth About How Innovative Companies and People Generate Great Ideas. Jossey-Bass.
4. Jain, R., Triandis, H. C., & Weick, C. W. (2010). Managing Research, Development and Innovation: Managing the Unmanageable. Wiley. Trott, P. (2011). Innovation Management and New Product Development (5th Edition). Prentice Hall.
5. Greenlick, M. (2012). Managing Research: The Cat-Herd's Toolkit. Inkwater Press. Jain, R., Triandis, H. C., & Weick, C. W. (2010). Managing Research, Development and Innovation: Managing the Unmanageable. Wiley.

Cycle of Innovation

[1] Sawyer, K. (2013). Zig Zag: The Surprising Path to Greater Creativity. Jossey-Bass.

[2] Cross contends that design thinking encompasses a set of processes including problem finding, problem definition, idea creation, solution finding, creative thinking, modelling, prototyping, testing and evaluating. These are very similar to the processes that we use in research to create knowledge about a scientific phenomenon as shown in this model. Cross, N. (2011). Design Thinking. Berg Publishers.

[3] Six Sigma is a process used in many manufacturing organizations to improve performance, to decrease variation and to eliminate defects. The term generally is used to indicate a well-controlled process. More information and training is available at https://asq.org/quality-resources/six-sigma. A history and general outline is available from Wikipedia at https://en.wikipedia.org/wiki/Six_Sigma. A summary is available at http://www.isixsigma.com/sixsigma/six_sigma.asp.

The Workplan

[1] Ben-Ari, M. (2005). Just a Theory: Exploring the Nature of Science. Prometheus Books.

Project Management

[1] Project Management Institute is the world's leading project management organization with over 500,000 members globally and over 300 local chapters internationally. For a full list of their services and publications, see https://www.pmi.org/

[2] Project Management Institute. (2017). A Guide to the Project Management Body of Knowledge (PMBOK Guide) Fifth Edition. Project Management Institute. Available at https://www.pmi.org/pmbok-guide-standards/foundational/pmbok.

[3] The website www2a.cdc.gov/cdcup/default.htm provides guides, templates and checklists for CDC projects that meet US federal requirements and standards.

[4] Lewis, J. (2010). Project Planning, Scheduling, and Control: The Ultimate Hands-On Guide to Bringing Projects in On Time and On Budget, Fifth Edition. McGraw-Hill. Lewis, J. P. (2004). Team-Based Project Management. Beard Books.

[5] Greenlick, M. (2012). Managing Research: The Cat-Herd's Toolkit. Inkwater Press. Jain, R., Triandis, H. C., & Weick, C. W. (2010). Managing Research, Development and Innovation: Managing the Unmanageable. Wiley.

[6] PMBOK® Guide and Standards are available at https://www.pmi.org/pmbok-guide-standards.

[7] Brown, K., & Hyer, N. (2009). Managing Projects: A Team-Based Approach with Student CD (McGraw-Hill/Irwin Series Operations and Decision Sciences). McGraw-Hill/Irwin.

8 To view the full report see https://www.standishgroup.com/sample_research_files/chaos_report_1994.pdf.

9 Broza, G. (2015). The Agile Mind-Set: Making Agile Processes Work. CreateSpace Independent Publishing Platform.Cobb, C. G. (2015). The Project Manager's Guide to Mastering Agile: Principles and Practices for an Adaptive Approach. Wiley. Layton, M. C., & Ostermiller, S. J. (2017). Agile Project Management for Dummies. For Dummies; Medinilla, Á. (2012). Agile Management: Leadership in an Agile Environment. Springer; O'Brien, H. (2015). Agile Project Management: A Quick Start Beginner's Guide to Mastering Agile Project Management. CreateSpace Independent Publishing Platform; Project Management Institute. (2017). Agile Practice Guide at https://www.pmi.org/pmbok-guide-standards.

10 Written in 2001 by a group calling "The Agile Alliance", the manifesto and principles are available at https://agilemanifesto.org/ and https://agilemanifesto.org/principles.html. This marked a change in the philosophy for managing projects in software development that has spread to many of types of creative development projects.

11 Training in agile project management is available from several consulting companies including Adventures in Agile (https://www.adventureswithagile.com/) that includes posts discussing the Declaration of Interdependence (https://www.adventureswithagile.com/2014/08/19/declaration-of-interdependence/) and the Agile Manifesto (https://www.adventureswithagile.com/2018/05/03/using-agile-principles-to-develop-company-culture-part-1-introduction/).

12 Virine, L., & Trumper, M. (2007). Project Decisions: The Art and Science. Berrett-Koehler Publishers. Wysocki, R. (2010). Adaptive Project Framework: Managing Complexity in the Face of Uncertainty. Addison-Wesley Professional.

13 4-H is a highly respected youth development organization, focused on rural regions of US and Canada. They provided me with my earliest training in teamwork. The four H's represent core values: Head: managing, thinking; Heart: relating, caring; Hands: giving, working; Health: being, living. More information is available at https://4-h-canada.ca/about

14 Cited from his NY Times article available at https://www.nytimes.com/2006/10/20/opinion/20greenehed.html

15 The full article on adaptive project management is available at https://www.projecttimes.com/articles/adaptive-project-management.html

16 DeCarlo, D. (2004). eXtreme Project Management: Using Leadership, Principles, and Tools to Deliver Value in the Face of Volatility. Jossey-Bass.

Research Projects are Different

1 Moore, S., & Shangraw, R. W. J. (2011). Managing Risk and Uncertainty in Large-Scale University Research Projects. Research Management Review, 18(2), 1.

2 Chalmers, A. F. (2013). What Is This Thing Called Science? Hackett Publishing Company. Kuhn, T. S. (2012). The Structure of Scientific Revolutions: 50th Anniversary Edition. University of Chicago Press. Mayo, D. G. (1996). Error and the Growth of Experimental Knowledge (Science and Its Conceptual

[3] Foundations series). University of Chicago Press.
[3] Brown, K. A. (1988). Inventors at Work: Interviews with 16 Notable American Inventors. Microsoft Press.
[4] Brown, K. A. (1988). Inventors at Work: Interviews with 16 Notable American Inventors. Microsoft Press.
[5] Ioannidis, J. P. (2005). Why most published research findings are false. PLoS Med, 2(8), e124.
[6] Cornell University provides an introduction to the various metrics used to measure research impact at http://guides.library.cornell.edu/impact.
[7] Eyre-Walker A., & Stoletzki N. (2013) The Assessment of Science: The Relative Merits of Post-Publication Review, the Impact Factor, and the Number of Citations. PLoS Biol 11(10): e1001675. https://doi.org/10.1371/journal.pbio.1001675
[8] From the article in the NY Times "Why We Make Bad Decisions" (Oct 2013) available at https://www.nytimes.com/2013/10/20/opinion/sunday/why-we-make-bad-decisions.html
[9] Janis, I. L. (1982). Groupthink: Psychological Studies of Policy Decisions and Fiascoes. Cengage Learning. Janis, I. L. (2014). Crucial Decisions. Free Press. Janis, I. L., & Mann, L. (1979). Decision Making: A Psychological Analysis of Conflict, Choice, and Commitment. Free Press.
[10] Greenlick, M. (2012). Managing Research: The Cat-Herd's Toolkit. Inkwater Press. Jain, R., Triandis, H. C., & Weick, C. W. (2010). Managing Research, Development and Innovation: Managing the Unmanageable. Wiley.
[11] DeCarlo, D. (2004). eXtreme Project Management: Using Leadership, Principles, and Tools to Deliver Value in the Face of Volatility. Jossey-Bass.
[12] In an article published by the Project Management Institute, Baratta explains the traditional triple constraint model and proposes improvements. Baratta, A. (2006). The triple constraint: a triple illusion. Paper presented at PMI® Global Congress 2006—North America, Seattle, WA. Newtown Square, PA: Project Management Institute. available at https://www.pmi.org/learning/library/triple-constraint-erroneous-useless-value-8024
[13] A list if products regulated by the US Food and Drug Administration is available at https://www.fda.gov/. The Occupational Safety and Health Administration provides guidance for laboratory safety available at https://www.osha.gov/Publications/laboratory/OSHA3404laboratory-safety-guidance.pdf. The National Institutes of Health also provides guidelines on biotechnology research, recombinant DNA and gene therapy available at https://osp.od.nih.gov/biotechnology/nih-guidelines/.
[14] Laws, regulations, compliance guidelines and policies are available at https://www2.epa.gov/laws-regulations
[15] The United States Department of Agriculture Animal and Plant Health Inspection Service regulates research involving animal health and welfare, biotechnology, pest and disease control, plant health and wildlife. Details are available at https://www.aphis.usda.gov/aphis/home
[16] The National Institutes of Health Clinical Center reviews some of the ethical requirements for clinical research at https://clinicalcenter.nih.gov/recruit/ethics.html. Oliver, P. (2010). The Student's Guide to Research Ethics. Open

17 University Press.
DeCarlo, D. (2004). eXtreme Project Management: Using Leadership, Principles, and Tools to Deliver Value in the Face of Volatility. Jossey-Bass.
18 Moore, S., & Shangraw, R. W. J. (2011). Managing Risk and Uncertainty in Large-Scale University Research Projects. Research Management Review, 18(2), 1. Pritchard, C. L. (2015). Risk Management: Concepts and Guidance, Fifth Edition. Auerbach Publications.
19 Murphy's Law is expressed in various humorous sayings with the theme that anything that can go wrong will go wrong.
20 https://en.wikiquote.org/wiki/Enrico_Fermi
21 The full article on adaptive project management is available at https://www.projecttimes.com/articles/adaptive-project-management.html

Creative Thinking

1 Sarooghi, H., Libaers, D., & Burkemper, A. (2015). Examining the relationship between creativity and innovation: A meta-analysis of organizational, cultural, and environmental factors. Journal of Business Venturing, 30(5), 714-731.
2 Greenlick, M. (2012). Managing Research: The Cat-Herd's Toolkit. Inkwater Press; Jain, R., Triandis, H. C., & Weick, C. W. (2010). Managing Research, Development and Innovation: Managing the Unmanageable. Wiley.
3 For a more detailed explanation see http://www.nwlink.com/~donclark/hrd/bloom.html. For learning and instruction strategies, see http://www.nwlink.com/~donclark/hrd/strategy.html. Anderson, L. W., Krathwohl, D. R., Airasian, P.W., Cruikshank, K. A., Mayer, R. E., Pintrich, P. R., Raths, J., & Wittrock, M. C. (2001). A Taxonomy for Learning, Teaching, and Assessing: A revision of Bloom's Taxonomy of Educational Objectives. Pearson, Allyn & Bacon; Krathwohl, D. R. (2002). A Revision of Bloom's Taxonomy: An Overview. Theory into Practice, 41(4), 212-218.
4 More detailed information on philosophy, logic and reasoning is available at https://plato.stanford.edu. The website at https://www.butte.edu/departments/cas/tipsheets/thinking/reasoning.html describes deductive, inductive, and abductive approaches to reasoning. Deductive reasoning gives a guaranteed conclusion. Inductive reasoning gives a conclusion that is merely likely. Abductive reasoning is taking your best shot. Most scientific models used in research projects are based on abductive logic. Abductive logic is discussed by Thagard, P., & Shelley, C. (1997). Abductive Reasoning: Logic, Visual Thinking, and Coherence. In M. L. Dalla Chiara, K. Doets, D. Mundici, & J. van Benthem (Eds.), Volume One of the Tenth International Congress of Logic, Methodology and Philosophy of Science, Florence, August 1995 (pp. 413-427). Dordrecht: Springer Netherlands.
5 https://foursightonline.com/
6 G.Pucci instructs on creative thinking styles in his course "The Creative Thinker's Toolkit" available at https://www.thegreatcourses.com/courses/the-creative-thinker-s-toolkit.html.
7 In addition to the many published books on creativity, numerous internet resources are available for each of the creativity tools mentioned in this book. Most of these creativity tools are described in Wikipedia and in videos on

YouTube. An internet search on Google will identify several additional sites for each of these tools, often including templates.

8. The American Society for Quality (ASQ) website presents many management process and creativity tools with an emphasis on high quality standards at https://asq.org/quality-resources/learn-about-quality

9. The Creating Minds website discusses how to be creative. Their goal is to provide real and useful principles, tools, articles and quotes about all matters around being creative and using creativity. A free book is available online at http://creatingminds.org/.

10. IdeaConnection provides information on several creative thinking and problem solving tools and links to other websites at https://www.ideaconnection.com/thinking-methods/.

11. Litemind contains posts and reference books on creativity at https://litemind.com/category/creativity/.

12. Mindtools website presents 29 creativity tools in addition to many other management guides at https://www.mindtools.com/pages/main/newMN_CT.htm. Some of the descriptions require membership.

13. Mycoted specialises in creativity and innovation. Their website provides a wiki of techniques at https://www.mycoted.com/Category:Creativity_Techniques

14. Divergent thinking is a process to generate many ideas in a spontaneous, free-flowing, "non-linear" manner in a short amount of time without judgement. Brainstorming is a form of divergent thinking. Convergent thinking is the opposite of divergent thinking. It seeks the correct answer from a list of options. The process is often used to select ideas created by divergent thinking by judging their relative merits. Convergent thinking is discussed in the next chapter.

15. More quotes from Linus Pauling are available at https://www.goodreads.com/author/quotes/52938.Linus_Pauling

16. Michalko, M. (2006). Thinkertoys: A Handbook of Creative-Thinking Techniques (2nd Edition). Ten Speed Press.

17. Silverstein, D., Samuel, P., & DeCarlo, N. (2012). The Innovator's Toolkit: 50+ Techniques for Predictable and Sustainable Organic Growth. Wiley.

18. Edward de Bono was an author of several books on creative thinking. De Bono, E. (1999). Six Thinking Hats. Back Bay Books; De Bono, E. (2015). Lateral Thinking: Creativity Step by Step. Harper Colophon. See Further Reading and his website for a list https://www.debono.com/.

19. Brainstorming is a generic term referring to many methods to create options by gathering a list of ideas spontaneously contributed by a group without judgment on those ideas. Further details are available at https://www.mindtools.com/brainstm.html, https://writingcenter.unc.edu/tips-and-tools/brainstorming/, http://www.decide-guide.com/brainstorm/ and many other creative thinking websites. Brain writing 6-3-5 is a variation on Brainstorming in which six people are given a form and asked to provide three ideas for solving a problem in five minutes. Further details are available at at https://business.tutsplus.com/tutorials/how-to-use-brainwriting-for-rapid-idea-generation--cms-26451and http://creatingminds.org/tools/brainwriting.htm. The Delphi Method is a variation of brainstorming that involves a consensus decision-making process using a series of questionnaires

20. to obtain expert opinions. For further details and support see http://armstrong.wharton.upenn.edu/delphi2/ and https://www.mindtools.com/pages/article/newTMC_95.htm. See the RAND corporation website for a list of scholarly articles at https://www.rand.org/topics/delphi-method.html.
20. Groupthink is the behaviour of a group of people who desire harmony and conformity and avoid conflict. The group often makes poor decisions and choices to reach a consensus decisions without critical evaluation of alternative viewpoints. The group suppresses dissenting viewpoints, either actively or passively, and restricts outside influences or reviews.
21. De Bono, E. (1999). Six Thinking Hats. Back Bay Books. More detail, and links to the book and training is available from the de Bono website at https://www.debono.com/six-thinking-hats-summary. Adam Sicinski discusses the process at https://blog.iqmatrix.com/six-thinking-hats. Mindtools provides guidance on the process at https://www.mindtools.com/pages/article/newTED_07.htm.
22. The Creating Minds website briefly discusses SCAMPER with examples at http://creatingminds.org/tools/scamper.htm. LiteMind gives a more detailed description at https://litemind.com/scamper/. Eberle, R. (1987). Scamper: Games for imagination development. D.O.K. Publishers; Eberle, R. (1997). Scamper On: More Creative Games and Activities for Imagination Development. Prufrock Press; Eberle, R. (2008). Scamper: Creative Games and Activities for Imagination Development. Prufrock Press.
23. CreatingMinds describes the attribute listing approach at http://creatingminds.org/tools/attribute_listing.htm.
24. A morphological matrix is a technique built on attribute listing to create, modify and improve something that is complex including products, technology, processes, hypotheses and scientific models. Innovation Management discusses how to use a morphological matrix to generate ideas at https://innovationmanagement.se/imtool-articles/how-to-using-a-morphological-matrix-to-generate-ideas/. Silverstein, D., Samuel, P., & DeCarlo, N. (2012). The Innovator's Toolkit: 50+ Techniques for Predictable and Sustainable Organic Growth. Wiley.
25. An example of a matrix diagram is given at http://www.syque.com/quality_tools/toolbook/Matrix/example.htm
26. This process combines the attributes or characteristics of different products, ideas, scientific models or hypotheses to create new ones. Mycoted presents an example using the Heuristic Ideation Technique at https://www.mycoted.com/Heuristic_Ideation_Technique. Dave Gray writes about the technique at https://gamestorming.com/heuristic-ideation-technique/.
27. The Stanford Encyclopedia of Philosophy website gives a detailed discussion of analogy and analogy reasoning at https://plato.stanford.edu/entries/reasoning-analogy/
28. Biomimicry is a type of borrow and modify technique that adapts nature's solutions to problems to solve our problems. The process of learning from and then emulating natures ingenious solutions to complex problems is described at AskNature at https://biomimicry.org/asknature/. Bioinspiration and Biomimetics publishes research articles at https://iopscience.iop.org/journal/1748-3190. Biomimicry provides consulting and training described at https://

29. McCullough, D. (2015). The Wright Brothers. Simon & Schuster.
30. Flatow, I. (1993). They All Laughed. From Light Bulbs to Lasers: The Fascinating Stories Behind the Great Inventions That Have Changed Our Lives. Harper Perennial.
31. Biomimicry Institute website is available at https://biomimicry.org/what-is-biomimicry/
32. Jain, R., Triandis, H. C., & Weick, C. W. (2010). Managing Research, Development and Innovation: Managing the Unmanageable. Wiley.
33. F. John Reh writes about ways to provide positive feedback at https://www.thebalancecareers.com/giving-positive-feedback-2275335. The CommunityToolBox website has several articles on requesting and receiving feedback from the general community at https://ctb.ku.edu/en/table-of-contents. G.Pucci instructs on ways to request and respond to feedback in his course "The Creative Thinker's Toolkit" available at https://www.thegreatcourses.com/courses/the-creative-thinker-s-toolkit.html.
34. Thomas J. DeLong writes about feedback in his article "Three Questions for Effective Feedback" in the Harvard Business Review at https://hbr.org/2011/08/three-questions-for-effective-feedback.
35. Buckingham, M., & Goodall, A. (2019). The Feedback Fallacy. Harvard Business Review. Mar 01, 2019. https://hbr.org/product/the-feedback-fallacy/R1902G-HCB-ENG?referral=03069.
36. If you chose to be negative, Scott Berkun also provides a list of responses that will kill ideas. You can use these approaches when you feel threatened by someone else's ideas and success. Available at https://scottberkun.com/2006/idea-killers-ways-to-stop-ideas/. Scott Berkun also presents some alternative ways to respond positively when new ideas are presented to you at https://scottberkun.com/2006/idea-starters-ways-to-grow-ideas/.
37. Gerrard Puccio. The Creative Thinkers Toolkit. Available through Great Courses at www.thegreatcourses.com/courses/the-creative-thinker-s-toolkit.html
38. Burkus, D. (2013). The Myths of Creativity: The Truth About How Innovative Companies and People Generate Great Ideas. Jossey-Bass.
39. Kahneman, D. (2011). Thinking, Fast and Slow. Farrar, Straus and Giroux.
40. Decide-guide provides a list of different decision-making guides that can help you with this process at http://www.decide-guide.com/.

Selection Criteria

1. The use of Pareto Analysis in a decision-making process is described on the decide-guide website at http://www.decide-guide.com/pareto-analysis/. WhatIs provides an overview at https://whatis.techtarget.com/definition/Pareto-principle.
2. Koch, R. (2005). The 80/20 Individual: How to Build on the 20% of what You Do Best. Crown Publishing Group; Koch, R. (2011). Living the 80/20 Way: Work Less, Worry Less, Succeed More, Enjoy More. Quercus; Koch, R. (2013). The

80/20 Principle and 92 Other Powerful Laws of Nature: The Science of Success. Quercus.

3. Investopedia describes how the Pareto Principle is applied in economics at https://www.investopedia.com/terms/p/paretoprinciple.asp. To learn how the Pareto Principle is applied in baseball, read the article at https://www.beyondtheboxscore.com/2010/6/4/1501048/applying-the-parento-principle-80. To learn how the Pareto Principle is applied to gambling and betting, read the article at https://www.pinnacle.com/en/betting-articles/Betting-Strategy/The-Pareto-Principle-of-Prediction/FS52BE6XD4ZJMSBQ

4. A process is considered to be Pareto optimized if a change in allocation of resources to one attribute makes at least one other attribute worse. As a more complex example, the Pareto efficiency principle has also been applied to explain the difference in codon usage across species, which has been optimized in each species for transcript cost and for translational efficiency by natural selection. Seward, E. A., & Kelly, S. (2018). Selection-driven cost-efficiency optimization of transcripts modulates gene evolutionary rate in bacteria. Genome Biol, 19(1), 102.

5. See the post by Margaret Rouse at https://whatis.techtarget.com/definition/PMI-plus-minus-interesting-retrospective.

6. De Bono, E. (2015). Lateral Thinking: Creativity Step by Step. Harper Colophon.

7. Decision Matrix is a technique to evaluate ideas or options using an objective scoring system including a set of criteria. Also known as Grid Analysis, Pugh Matrix Analysis, Problem matrix, Opportunity analysis, Criteria rating form, Multi-Attribute Utility Theory and many other names. American Society for Quality asq.org/learn-about-quality/decision-making-tools/overview/decision-matrix.html. MindTools www.mindtools.com/pages/article/newTED_03.htm. University of Notre Dame blogs.nd.edu/jlugo/2012/09/24/pugh-method-how-to-decide-between-different-designs/.

8. Paired or Pairwise Comparison Analysis (PCA) is described by Continuous Improvement Toolkit in a slideshow at https://www.slideshare.net/dsaadeddin/paired-comparison-analysis-53581809. Strategy Works provides an example at https://www.strategyworks.co.za/2009/11/paired-comparison-analysis/. Variations of this approach are discussed with examples at CreatingMinds http://www.syque.com/quality_tools/toolbook/Priority/vary.htm. A more intensive comparison is Potentially All Pairwise RanKings of all possible Alternatives (PAPRIKA). This method for multi-criteria decision-making also uses the decision-makers' preferences as expressed using pairwise rankings of alternatives. 100Minds provides commercial software for this decision making process at https://www.1000minds.com/about/paprika.

9. Analytical Hierarchy Process is a technique to determine your relative preferences in a list of complex options, when considering multiple, potentially conflicting criteria. AHP is useful to compare ideas, problems, objectives, risks, hypotheses, experimental design and many other choices in a research project. Saaty, T. L. (2012). Decision Making for Leaders: The Analytic Hierarchy Process for Decisions in a Complex World. RWS Publications. MindTools provides a description at https://www.mindtools.com/pages/article/newTED_88.htm. K Buruamkaewu shows a presentation on AHP using Excel at https://www.slideshare.net/rabbittrix/how-to-do-ahp-analysis-in-excel.

Review Panels

1. Keeney, S., McKenna, H., & Hasson, F. (2011). The Delphi Technique in Nursing and Health Research. Wiley-Blackwell.
2. Finkelstein, S. (2019). Don't Be Blinded by Your Own Expertise. Harvard Business Review. May–June 2019, 153–158. https://hbr.org/2019/05/dont-be-blinded-by-your-own-expertise.
3. Kuhn, T. S. (2012). The Structure of Scientific Revolutions: 50th Anniversary Edition. University of Chicago Press.
4. Stephen Malker gives a summary of lessons learned from Collin Powell at http://www.stephenwalker.com/notes/colin-powell-on-leadership/.
5. Scott London discusses "Thinking Together: The Power of Deliberative Dialogue" at http://scott.london/reports/dialogue.html.

Group Decision-making

1. See the decide-guide website for several article on using consensus decision making at http://www.decide-guide.com/consensus-decision-making/.
2. Some project management manuals require at least 70% support, but that is a poorly defined measurement.
3. Lewis, J. (2010). Project Planning, Scheduling, and Control: The Ultimate Hands-On Guide to Bringing Projects in On Time and On Budget, Fifth Edition. McGraw-Hill; Lewis, J. P. (2004). Team-Based Project Management. Beard Books.
4. Janis, I. L. (1972). Victims of Groupthink: A psychological study of foreign-policy decisions and fiascoes. Houghton Mifflin Company. Janis, I. L. (1982). Groupthink: Psychological Studies of Policy Decisions and Fiascoes. Cengage Learning.
5. Greenlick, M. (2012). Managing Research: The Cat-Herd's Toolkit. Inkwater Press; Jain, R., Triandis, H. C., & Weick, C. W. (2010). Managing Research, Development and Innovation: Managing the Unmanageable. Wiley.

The Problem

1. The original eight Millennium Development Goals ranged from halving extreme poverty to halting the spread of HIV/AIDS and providing universal primary education by the target date of 2015 and was widely viewed as being successful because they galvanized unprecedented efforts to meet the needs of the world's poorest. This agreement was updated with the 2030 Agenda for Sustainable Development. The new agenda calls on countries to begin efforts to achieve 17 goals. For more details and reports on the achievement of the eight goals, see https://www.un.org/millenniumgoals/.
2. The Grand Challenges initiatives foster innovation to solve key health and development problems. Grants are provided by Bill & Melinda Gates Foundation other salon or in partnership with other groups. For details, see https://gcgh.grandchallenges.org/about.
3. The survey "This is what millennials want in 2018" is available at https://

4. Statistics were taken from https://seer.cancer.gov/statfacts/html/breast.html.
5. Mark Levy's article "A Problem Well-stated is Half-solved" gives further advice on wording a problem. Read his article at http://www.levyinnovation.com/a-problem-well-stated-is-half-solved/.
6. Jain, R., Triandis, H. C., & Weick, C. W. (2010). Managing Research, Development and Innovation: Managing the Unmanageable. Wiley.
7. Originally called the 5-Whys but other questions may be used in place of why. The Asian Development Bank provides a free pdf description at https://www.adb.org/publications/five-whys-technique. An Introduction to 5-why is available at https://www.bulsuk.com/2009/03/5-why-finding-root-causes.html. Wikipedia gives history and examples at https://en.wikipedia.org/wiki/Five_whys. A spreadsheet approach using Excel is available at https://www.bulsuk.com/2009/07/5-why-analysis-using-table.html.
8. Finkelstein, S. (2019). Don't Be Blinded by Your Own Expertise. Harvard Business Review. May–June 2019, 153–158. https://hbr.org/2019/05/dont-be-blinded-by-your-own-expertise.
9. Other quotes from Albert Einstein are available at http://www.gurteen.com/gurteen/gurteen.nsf/id/X00001A96/.

Observation Projects

1. This statement is often quoted as a motivation for innovation but perhaps Ford never actually said these exact words as cited in the Harvard Business Review article by Patrick Vlaskovits at https://hbr.org/2011/08/henry-ford-never-said-the-fast
2. Chalmers, A. F. (2013). What Is This Thing Called Science? Hackett Publishing Company, Inc.; Mayo, D. G. (1996). Error and the Growth of Experimental Knowledge. University Of Chicago Press.
3. For example, GEO is a genomics data repository of array- and sequence-based data. Tools are also provided to help users query and download experiments and curated gene expression profiles available at https://www.ncbi.nlm.nih.gov/geo/. The journal Nature provides a list of its recommended repositories at https://www.nature.com/sdata/policies/repositories.

The Vision

1. A paradigm shift according to Kuhn is a fundamental change in the basic concepts, theories and scientific models of a scientific discipline. Kuhn, T. S. (2012). The Structure of Scientific Revolutions: 50th Anniversary Edition. University of Chicago Press.
2. Schrage, M. (2012). Who Do You Want Your Customers to Become? Harvard Business Review Press.
3. A thought experiment considers the consequences of a hypothesis, theory, or

4. principle without conducting a physical experiment. Most notable are the thought experiments of Albert Einstein. It may not be practical, legal or ethical to actually conduct the experiment. A list of examples is given in Wikipedia https://en.wikipedia.org/wiki/Thought_experiment. A list of publications in philosophy is maintained at https://philpapers.org/browse/thought-experiments and in the philosophy of science at http://philsci-archive.pitt.edu/view/subjects/thought-experiments.html.
4. Jain, R., Triandis, H. C., & Weick, C. W. (2010). Managing Research, Development and Innovation: Managing the Unmanageable. Wiley.
5. CRISPR-Cas9 is a genetic engineering technology that can be used to make nucleotide changes in DNA and has the potential to repair mutations leading to cancer. Doudna, J. A., & Sternberg, S. H. (2018). A Crack in Creation: Gene Editing and the Unthinkable Power to Control Evolution. Mariner Books.
6. De Bono, E. (1999). Six Thinking Hats. Back Bay Books.
7. MindTools describes "Force Field Analysis Analyzing the Pressures For and Against Change" and give step by step instructions at https://www.mindtools.com/pages/article/newTED_06.htm.
8. Mark Connelly describes force field analysis according to Kurt Lewin at https://www.change-management-coach.com/force-field-analysis.html. Jim Riley discusses "Lewin's Force Field Model (Change Management)" at https://www.tutor2u.net/business/reference/models-of-change-management-lewins-force-field-model.

Literature Reviews

1. Salkind, N. J. (2011). 100 Questions (and Answers) About Research Methods. SAGE Publications, Inc.
2. Karl Popper was a very influential philosopher of science who advocated that a theory in the empirical sciences can never be proven, it can only be falsified.
3. Chalmers, A. F. (2013). What Is This Thing Called Science? Hackett Publishing Company, Inc.
4. Ibid
5. Sean Covey Quotes. (n.d.). BrainyQuote.com. Retrieved August 31, 2019, from BrainyQuote.com Web site: https://www.brainyquote.com/quotes/sean_covey_657208
6. Some commercial software for managing scientific references includes Endnote https://endnote.com/, Reference Manager https://mendeley.com/ and Bookends https://www.sonnysoftware.com/bookends/bookends.html.
7. Read the article Baylor College of Medicine at https://www.bcm.edu/news/technology/knowledge-integration-toolkit-advances.
8. Watson is IBM's suite of enterprise-ready AI services, applications, and tooling. For details on past and present projects, see their website at https://www.ibm.com/watson.
9. Spangler, S., Wilkins, A. D., Bachman, B. J., Nagarajan, M., Dayaram, T., Haas, P., . . . Lichtarge, O. (2014). Automated Hypothesis Generation Based on Mining Scientific Literature Proceedings of the 20th ACM SIGKDD International Conference on Knowledge Discovery and Data Mining. Proceedings from KDD '14, New York, New York, USA New York, NY, USA.

10   A meta-analysis is a statistical analysis that combines the data from multiple scientific studies. More details are available at https://en.wikipedia.org/wiki/Meta-analysis. Courses and software are available from several sources including https://www.meta-analysis.com/pages/why_do.php?cart=BFCU3172093

11   Wikipedia gives an overview of the technology at https://en.wikipedia.org/wiki/DNA_microarray. GEO is a genomics data repository of array- and sequence-based data. Tools are also provided to help users query and download experiments and curated gene expression profiles available at https://www.ncbi.nlm.nih.gov/geo/. The journal Nature provides a list of its recommended repositories at https://www.nature.com/sdata/policies/repositories.

12   The Preferred Reporting Items for Systematic Reviews and Meta-Analyses (PRISMA) website is http://www.prisma-statement.org/. PRISMA aims to help authors improve the reporting of systematic reviews and meta-analyses. Moher, D., Liberati, A., Tetzlaff, J., & Altman, D. G. (2009). Preferred reporting items for systematic reviews and meta-analyses: the PRISMA statement. PLoS Med, 6, e1000097. Other reviews are found in the Current Opinion series at https://www.elsevier.com/life-sciences/journals/core/current-opinion; The Trends in... series at https://www.cell.com/; The Annual Review of ... series at https://www.annualreviews.org/action/showPublications; and other peer-reviewed journals.

13   To read about the background and for examples, see the article on Quality Progress at http://asq.org/quality-progress/2007/08/problem-solving/have-you-adequately-defined-your-situation.html. Lewis, J. (2010). Project Planning, Scheduling, and Control: The Ultimate Hands-On Guide to Bringing Projects in On Time and On Budget, Fifth Edition. McGraw-Hill.

14   Silverstein, D., Samuel, P., & DeCarlo, N. (2012). The Innovator's Toolkit: 50+ Techniques for Predictable and Sustainable Organic Growth. Wiley.

15   Ishikawa diagrams are also called fishbone diagrams or cause-and-effect diagrams. A graphical method to identify the root causes of a specific event. ASQ https://asq.org/quality-resources/fishbone. Karn Bulsuk's post https://www.bulsuk.com/2009/08/using-fishbone-diagram-to-perform-5-why.html. Business Excellence https://www.quality-assurance-solutions.com/Fish-Bone.html. Templates for Word or Excel from iSixSigma https://www.isixsigma.com/tools-templates/cause-effect/cause-and-effect-aka-fishbone-diagram/. Another template in Excel format is available at https://asq.org/quality-resources/fishbone. Silverstein, D., Samuel, P., & DeCarlo, N. (2012). The Innovator's Toolkit: 50+ Techniques for Predictable and Sustainable Organic Growth. Wiley.

16   A dendrogram is a branching diagram showing a hierarchical clustering produced by a computational analysis, for example the clustering of genes based on sequence similarity. More information is available at https://ncss-wpengine.netdna-ssl.com/wp-content/themes/ncss/pdf/Procedures/NCSS/Hierarchical_Clustering-Dendrograms.pdf and https://www.displayr.com/what-is-dendrogram/. A phylogenetic tree is a branching diagram showing the evolutionary relationships among various biological species.

17   An Interrelationship Diagram is also called relations diagram or network

diagram that can serve as a tool for root cause identification. It is used as a summary prepared from a literature review to better understand potential cause-effect relationships. More information is available from ASQ at https://asq.org/quality-resources/relations-diagram. SmartDraw https://www.smartdraw.com/interrelationship-diagram/. SkyMark https://www.skymark.com/resources/tools/relations_diagram.asp. BusinessExcellence https://www.quality-assurance-solutions.com/Interrelationship-diagram.html.

[18] Mind mapping will visually organize information and relationships in a tree-like structure showing relationships. Several forms of software are available, or diagrams can be free-hand. Further details are available at iMindMap https://www.ayoa.com/how-to-mind-map/?utm_medium=301&utm_source=imindmap.com, Mind Mapping https://www.mindmapping.com/index.php. MindTools https://www.mindtools.com/pages/article/newISS_01.htm and decide-guide http://www.decide-guide.com/mind-mapping/.

[19] Wikipedia lists and compares mindmap software for different platforms at https://en.wikipedia.org/wiki/List_of_concept-_and_mind-mapping_software.

[20] For examples in healthcare see https://www.pritikin.com/what-is-reverse-causation.

The Scientific Model

[1] ScienceOrNot? website will help you separate real science from nonsense that's masquerading as science at https://scienceornot.net/. There are examples showing how it's done. There is also a blog with relevant articles.

[2] Chalmers, A. F. (2013). What Is This Thing Called Science? Hackett Publishing Company, Inc.;Copi, I. M., Cohen, C., & McMahon, K. (2010). Introduction to Logic (14th Edition). Pearson; Kuhn, T. S. (2012). The Structure of Scientific Revolutions: 50th Anniversary Edition. University of Chicago Press; Rosenberg, A. (2011). Philosophy of Science: A Contemporary Introduction. Routledge.

[3] Chalmers, A. F. (2013). What Is This Thing Called Science? Hackett Publishing Company, Inc.

[4] More detailed information on philosophy, logic and reasoning is available at https://plato.stanford.edu. The website at https://www.butte.edu/departments/cas/tipsheets/thinking/reasoning.html describes deductive, inductive, and abductive approaches to reasoning. Deductive reasoning gives a guaranteed conclusion. Inductive reasoning gives a conclusion that is merely likely. Abductive reasoning is taking your best shot. Most scientific models used in research projects are based on abductive logic. Abductive logic is discussed by Thagard, P., & Shelley, C. (1997). Abductive Reasoning: Logic, Visual Thinking, and Coherence. In M. L. Dalla Chiara, K. Doets, D. Mundici, & J. van Benthem (Eds.), Volume One of the Tenth International Congress of Logic, Methodology and Philosophy of Science, Florence, August 1995 (pp. 413-427). Dordrecht: Springer Netherlands.

[5] Further explanation by Gershenfeld and related topics is available at https://www.edge.org/response-detail/10395. A more detailed discussion of this quote

6. Popper, K. (2002). The Logic of Scientific Discovery. Routledge.
7. Read Sam Kean's article in The Atlantic at https://www.theatlantic.com/science/archive/2017/12/trofim-lysenko-soviet-union-russia/548786/.
8. A thought experiment considers the consequences of a hypothesis, theory, or principle without conducting a physical experiment. Most notable are the thought experiments of Albert Einstein. It may not be practical, legal or ethical to actually conduct the experiment. A list of examples is given in Wikipedia https://en.wikipedia.org/wiki/Thought_experiment. A list of publications in philosophy is maintained at https://philpapers.org/browse/thought-experiments and in the philosophy of science at http://philsci-archive.pitt.edu/view/subjects/thought-experiments.html.
9. A more complete discussion of the Principle of Parsimony is available at https://science.howstuffworks.com/innovation/scientific-experiments/occams-razor.htm.
10. Karl Popper was a very influential philosopher of science who advocated that a theory in the empirical sciences can never be proven, it can only be falsified. Popper, K. (2002). The Logic of Scientific Discovery. Routledge.
11. Kuhn, T. S. (2012). The Structure of Scientific Revolutions: 50th Anniversary Edition. University of Chicago Press.
12. To read more about this quote see https://www.edge.org/response-detail/25332.
13. A complete description is available at https://plato.stanford.edu/entries/scientific-underdetermination/. A YouTube video is available at https://www.youtube.com/watch?v=-klqI4d_wbY.
14. Chalmers, A. F. (2013). What Is This Thing Called Science? Hackett Publishing Company, Inc.
15. For an overview, read the article in Wikipedia at https://en.wikipedia.org/wiki/Argument.

Modeling Projects

1. More detailed information on philosophy, logic and reasoning is available at https://plato.stanford.edu. The website at https://www.butte.edu/departments/cas/tipsheets/thinking/reasoning.html describes deductive, inductive, and abductive approaches to reasoning. Deductive reasoning gives a guaranteed conclusion. Inductive reasoning gives a conclusion that is merely likely. Abductive reasoning is taking your best shot. Most scientific models used in research projects are based on abductive logic.
2. A thought experiment considers the consequences of a hypothesis, theory, or principle without conducting a physical experiment. Most notable are the thought experiments of Albert Einstein. It may not be practical, legal or ethical to actually conduct the experiment. A list of examples is given in Wikipedia https://en.wikipedia.org/wiki/Thought_experiment. A list of publications in philosophy is maintained at https://philpapers.org/browse/thought-experiments and in the philosophy of science at http://philsci-archive.pitt.edu/view/subjects/thought-experiments.html.

³ Ben-Ari, M. (2005). Just a Theory: Exploring the Nature of Science. Prometheus Books.
⁴ Divergent thinking is a process to generate many ideas in a spontaneous, free-flowing, "non-linear" manner in a short amount of time without judgement. Brainstorming is a form of divergent thinking.
⁵ Convergent thinking is the opposite of divergent thinking. It seeks the correct answer from a list of options. The process is often used to select ideas created by divergent thinking by judging their relative merits.
⁶ Newton, D. E. (2014). GMO Food: A Reference Handbook. ABC-CLIO.
⁷ Kuhn, T. S. (2012). The Structure of Scientific Revolutions: 50th Anniversary Edition. University of Chicago Press.
⁸ Karl Popper was a very influential philosopher of science who advocated that a theory in the empirical sciences can never be proven, it can only be falsified. Popper, K. (2002). The Logic of Scientific Discovery. Routledge.

The Objective

¹ To read more about this quote see the article on philosiblog at https://philosiblog.com/2011/07/13/if-you-dont-know-where-youre-going/.
² To read more about this quote see the article on philosiblog at https://philosiblog.com/2014/03/16/no-matter-what-our-motivation-may-be-if-we-are-not-realistic-we-will-not-fulfill-our-goal/.

The Research Concept

¹ Sawyer, K. (2013). Zig Zag: The Surprising Path to Greater Creativity. Jossey-Bass.
² Investopedia describes SWOT analysis at https://www.investopedia.com/terms/s/swot.asp. Other articles are available at BusinessNewsDaily https://www.businessnewsdaily.com/4245-swot-analysis.html. Community ToolBox https://ctb.ku.edu/en/table-of-contents/assessment/assessing-community-needs-and-resources/swot-analysis/main. MindTools https://www.mindtools.com/pages/article/newTMC_05.htm. Decide-guide http://www.decide-guide.com/swot/
³ Read the article by Lisa Furgison "SWOT Analysis Step 5: Developing Actionable Strategies" available at https://articles.bplans.com/swot-analysis-challenge-day-5-turning-swot-analysis-actionable-strategies/.
⁴ MindTools describes "Force Field Analysis Analyzing the Pressures For and Against Change" and give step by step instructions at https://www.mindtools.com/pages/article/newTED_06.htm.
⁵ DeCarlo, D. (2004). eXtreme Project Management: Using Leadership, Principles, and Tools to Deliver Value in the Face of Volatility. Jossey-Bass.
⁶ Ibid

The Research Proposal

Understanding Sponsors

1. UNESCO Institute of Statistics reports that as part of the Sustainable Development Goals, countries have pledged to substantially increase public and private R&D spending as well as the number of researchers by 2030. More information and recent data are available at http://uis.unesco.org/apps/visualisations/research-and-development-spending/. The World Bank also tracks global spending on R&D at https://data.worldbank.org/indicator/GB.XPD.RSDV.GD.ZS.
2. Report from the National Science Board for 2018 is available at https://www.nsf.gov/statistics/2018/nsb20181/. Reports are updated annually.
3. From the National Science Board news release on "Science and Engineering Indicators 2010" available at https://www.nsf.gov/nsb/news/news_summ.jsp?org=NSB&cntn_id=116238. The release indicates that "In terms of R&D expenditures as a share of economic output, while Japan has surpassed the U.S. for quite some time, South Korea is now in the lead--ahead of the U.S. and Japan."
4. The OECD data can be accessed at https://stats.oecd.org/Index.aspx?DataSetCode=MSTI_PUB.
5. The report on the Global Innovation Index from 2019 can be accessed at https://www.globalinnovationindex.org/homehttps://www.globalinnovationindex.org/home. Reports are updated annually.
6. Bloomberg Innovation Index from 2015 is available at https://www.bloomberg.com/graphics/2015-innovative-countries/https://www.bloomberg.com/graphics/2015-innovative-countries/
7. The report based on 2013 data is available at https://www.conferenceboard.ca/hcp/details/innovation.aspx?AspxAutoDetectCookieSupport=1.
8. The 2020 strategic plan for NSERC is available at http://www.nserc-crsng.gc.ca/NSERC-CRSNG/NSERC2020-CRSNG2020/NSERC2020-2020CRSNG_eng.pdfhttp://www.nserc-crsng.gc.ca/NSERC-CRSNG/NSERC2020-CRSNG2020/NSERC2020-2020CRSNG_eng.pdf
9. UC Berkley explains who pays for science at https://undsci.berkeley.edu/article/0_0_0/who_pays. American Association for the Advancement of Science reports on the status of research funding in 2014 at https://www.aaas.org/news/rd-fy-2014-omnibus-big-picture and reports on the status of manufacturing R&D in the 2020 budget at https://www.aaas.org/news/fy-2020-budget-request-manufacturing-rd.
10. As reported by The Guardian at https://www.theguardian.com/science/political-science/2013/jun/26/spending-review-2013-science-innovation.
11. Arora, A., Belenzon, S., & Patacconi, A. (2015). Killing the Golden Goose? The Decline of Science in Corporate R&D. National Bureau of Economic Research Working Paper Series, No. 20902.
12. Ibid
13. Ibid
14. Ibid
15. Further details on the calculation of ROI are available at https://en.wikipedia.org/wiki/Return_on_investment and https://

16 www.investopedia.com/terms/r/returnoninvestment.asp.
17 Jain, R., Triandis, H. C., & Weick, C. W. (2010). Managing Research, Development and Innovation: Managing the Unmanageable. Wiley.
18 Details are discussed at https://en.wikipedia.org/wiki/Net_present_value and https://www.investopedia.com/terms/n/npv.asp.
19 Lavallo, D., & Kahneman, D. (2003). Delusions of Success: How Optimism Undermines Executives' Decisions. Harvard Business Review, 1-10.
20 The calculator is available at https://www.investopedia.com/calculator/netpresentvalue.aspx.
21 Wikipedia provides several article that give an overview of Net Present Value and discount rates.

Read the article at https://www.investopedia.com/ask/answers/06/npvdisadvantages.asp.

Selling the Research Proposal

1 From the TED talk "Why Should We Trust Scientists?" available at https://www.ted.com/talks/naomi_oreskes_why_we_should_believe_in_science/transcript. For further insights Oreskes, N. (2019). Why Trust Science? Princeton University Press.
2 The full article is available on LinkedIn at https://www.linkedin.com/pulse/creative-problem-solving-tool-success-zones-download-peter-tinker/.

Further Reading

Innovation

Bell, A., Chetty, R., Jaravel, X., Petkova, N., & Van Reenen, J. (2017). Who Becomes an Inventor in America? The Importance of Exposure to Innovation. http://www.equality-of-opportunity.org/assets/documents/inventors_paper.pdf

Berkun, S. (2010). *The Myths of Innovation*. O'Reilly Media.

Blank, S., & Dorf, B. (2012). *The Startup Owner's Manual: The Step-By-Step Guide for Building a Great Company*. K & S Ranch.

Brown, K. A. (1988). *Inventors at Work: Interviews with 16 Notable American Inventors*. Microsoft Press.

Burkus, D. (2013). *The Myths of Creativity: The Truth About How Innovative Companies and People Generate Great Ideas*. Jossey-Bass.

Carlson, C. R., & Wilmot, W. W. (2006). *Innovation: The Five Disciplines for Creating What Customers Want*. Crown Business.

Davila, T., Epstein, M., & Shelton, R. (2012). *Making Innovation Work: How to Manage It, Measure It, and Profit from It, Updated Edition*. Pearson FT Press.

Drucker, P. F. (1985). *Innovation and Entrepreneurship: Practice and Principles*. Harpercollins.

Edgett, S. (2014). People: A Key to Innovation Capability. *European Business Review*, 10-12.

Flatow, I. (1993). *They All Laughed. From Light Bulbs to Lasers: The Fascinating Stories Behind the Great Inventions That Have Changed Our Lives*. Harper Perennial.

Johnson, S. (2015). *How We Got to Now: Six Innovations That Made the Modern World*. Riverhead Books.

Lencioni, P. (2012). *The Advantage*. John Wiley & Sons.

Ries, E. (2011). *The Lean Startup: How Today's Entrepreneurs Use Continuous Innovation to Create Radically Successful Businesses*. Crown Business.

Sarooghi, H., Libaers, D., & Burkemper, A. (2015). Examining the relationship between creativity and innovation: A meta-analysis of organizational, cultural, and environmental factors. *Journal of Business Venturing, 30*(5), 714-731.

Trott, P. (2011). *Innovation Management and New Product Development (5th Edition)*. Prentice Hall.

Youn, H., Strumsky, D., Bettencourt, L. M. A., & Lobo, J. (2015). Invention as a combinatorial process: evidence from US patents. *Journal of The Royal Society Interface, 12*(106), DOI: 10.1098/rsif.2015.0272.

Project Management Manuals

Brown, J. T. (2014). *The Handbook of Program Management: How to Facilitate Project Success with Optimal Program Management, Second Edition*. McGraw-Hill Education.

Brown, K., & Hyer, N. (2009). *Managing Projects: A Team-Based Approach*. McGraw-

Hill/Irwin.

Broza, G. (2015). *The Agile Mind-Set: Making Agile Processes Work*. CreateSpace Independent Publishing Platform.

Campbell, G. M. (2014). *Project Management, Sixth Edition (Idiot's Guides)*. Alpha.

Cleland, D. I., & Ireland, L. (1999). *Project Manager's Portable Handbook*. McGraw-Hill Publishing Co.

Cobb, C. G. (2015). *The Project Manager's Guide to Mastering Agile: Principles and Practices for an Adaptive Approach*. John Wiley & Sons.

Craig, J. C. (2012). *Project Management Lite: Just Enough to Get the Job Done.Nothing More*. CreateSpace Independent Publishing Platform.

DeCarlo, D. (2004). *eXtreme Project Management: Using Leadership, Principles, and Tools to Deliver Value in the Face of Volatility*. Jossey-Bass.

Devaux, S. A. (1999). *Total Project Control: A Manager's Guide to Integrated Project Planning, Measuring, and Tracking*. Wiley.

Hass, K. B. (2008). *Managing Complex Projects: A New Model*. Berrett-Koehler Publishers.

Jordan, A. (2013). *Risk Management for Project Driven Organizations: A Strategic Guide to Portfolio, Program and PMO Success*. J Ross Publishing.

Kerzner, H. (2011). *Using the Project Management Maturity Model: Strategic Planning for Project Management*. Google eBook.

Kerzner, H. (2014). *Project Management Best Practices: Achieving Global Excellence*. Wiley.

Kerzner, H. (2017). *Project Management: A Systems Approach to Planning, Scheduling, and Controlling*. Wiley.

Kerzner, H. R., & Belack, C. (2010). *Managing Complex Projects*. Wiley.

Klastorin, T. (2011). *Project Management: Tools and Trade-offs*. Pearson Learning Solutions.

Layton, M. C., & Ostermiller, S. J. (2017). *Agile Project Management For Dummies*. For Dummies.

Lewis, J. (2010). *Project Planning, Scheduling, and Control: The Ultimate Hands-On Guide to Bringing Projects in On Time and On Budget, Fifth Edition*. McGraw-Hill.

Lewis, J. P. (2004). *Team-Based Project Management*. Beard Books.

Medinilla, Á. (2012). *Agile Management: Leadership in an Agile Environment*. Springer.

O'Brien, H. (2015). *Agile Project Management: A Quick Start Beginner's Guide To Mastering Agile Project Management*. CreateSpace Independent Publishing Platform.

Pritchard, C. L. (2015). *Risk Management: Concepts and Guidance, Fifth Edition*. Auerbach Publications.

Project Management Institute. (2017). *Agile Practice Guide*. Project Management Institute. https://www.pmi.org/pmbok-guide-standards/practice-guides/agile

Project Management Institute (2017). *A Guide to the Project Management Body of Knowledge (PMBOK Guide) Sixth Edition*. https://www.pmi.org/pmbok-guide-standards/foundational

Project Management Institute (2017). *The Standard for Portfolio Management-Fourth Edition.* https://www.pmi.org/pmbok-guide-standards/foundational

Project Management Institute (2017). *The Standard for Program Management - Fourth Edition.* https://www.pmi.org/pmbok-guide-standards/foundational

Project Management Institute (2017). *PMI Lexicon of Project Management Terms version 3.2.* https://www.pmi.org/pmbok-guide-standards/lexicon

Project Management Institute (2013). Organizational Project Management Maturity Model (OPM3®) – Third Edition. https://www.pmi.org/pmbok-guide-standards/foundational

Taylor, P. (2010). *The Lazy Project Manager: How to be twice as productive and still leave the office early.* Infinite Ideas.

Virine, L., & Trumper, M. (2007). *Project Decisions: The Art and Science.* Berrett-Koehler Publishers.

Wong, Z. (2007). *Human Factors in Project Management: Concepts, Tools, and Techniques for Inspiring Teamwork and Motivation.* John Wiley & Sons.

Wysocki, R. (2010). *Adaptive Project Framework: Managing Complexity in the Face of Uncertainty.* Addison-Wesley Professional.

Wysocki, R. K. (2013). *Effective Project Management: Traditional, Agile, Extreme.* Wiley.

PHILOSOPHY AND SCIENTIFIC METHOD

Asimov, I. (1994). *Asimov's Chronology of Science & Discovery: Updated and Illustrated.* Harper Collins.

Bell, J. (2010). *Doing Your Research Project.* Open University Press.

Ben-Ari, M. (2005). *Just a Theory: Exploring the Nature of Science.* Prometheus Books.

Carey, S. S. (2011). *A Beginner's Guide to Scientific Method.* Cengage Learning.

Chalmers, A. F. (2013). *What Is This Thing Called Science?* Hackett Publishing Company, Inc.

Derry, G. N. (1999). *What Science Is and How It Works.* Princeton University Press.

Gauch, H. G. J. (2012). *Scientific Method in Brief.* Cambridge University Press.

Gimbel, S. (2011). *Exploring the Scientific Method: Cases and Questions.* University Of Chicago Press.

Godfrey-smith, P. (2007). *Theory and Reality - An Introduction to the Philosophy of Science.* University of Chicago Press.

Goodwin, C. J., & Goodwin, K. A. (2012). *Research in Psychology: Methods and Design.* Wiley.

Greenlick, M. (2012). *Managing Research: The Cat-Herd's Toolkit.* Inkwater Press.

Howson, C., & Urbach, P. (1989). *Scientific Reasoning: The Bayesian Approach.* Open Court Pub Co.

Jain, R., Triandis, H. C., & Weick, C. W. (2010). *Managing Research, Development and Innovation: Managing the Unmanageable.* Wiley.

Kuhn, T. S. (2012). *The Structure of Scientific Revolutions: 50th Anniversary Edition.*

University of Chicago Press.

Mayo, D. G. (1996). *Error and the Growth of Experimental Knowledge*. University Of Chicago Press.

McComas, W. F. (2000). *The Nature of Science in Science Education: Rationales and Strategies*. Springer.

Nola, R., & Sankey, H. (2007). *Theories of Scientific Method: An Introduction*. McGill Queens University Press.

Okasha, S. (2016). *Philosophy of Science: Very Short Introduction*. OUP Oxford.

Oreskes, N. (2019). *Why Trust Science?* Princeton University Press.

Popper, K. (2002). *The Logic of Scientific Discovery*. Routledge.

Rosenberg, A. (2011). *Philosophy of Science: A Contemporary Introduction*. Routledge.

Rutherford, F. J., & Ahlgren, A. (1991). *Science for All Americans*. Oxford University Press.

Salkind, N. J. (2011). *100 Questions (and Answers) About Research Methods*. SAGE Publications, Inc.

CREATIVE THINKING AND PROBLEM SOLVING

Cox, D. (2013). *Creative Thinking For Dummies*. For Dummies.

Crawford, R. P. (2012). *The Techniques Of Creative Thinking: How To Use Your Ideas To Achieve Success*. Literary Licensing, LLC.

De Bono, E. (1999). *Six Thinking Hats*. Back Bay Books.

De Bono, E. (2015). *Lateral Thinking: Creativity Step by Step*. Harper Colophon.

DeLong, T. J. (2011). Three Questions for Effective Feedback. *Harvard Business Review*.

Eberle, R. (1987). *Scamper: Games for imagination development*. D.O.K. Publishers.

Eberle, R. (1997). *Scamper On: More Creative Games and Activities for Imagination Development*. Prufrock Press.

Eberle, R. (2008). *Scamper: Creative Games and Activities for Imagination Development*. Prufrock Press.

Fantin, I. (2014). *Applied Problem Solving: Method, Applications, Root Causes, Countermeasures, Poka-Yoke and A3*. CreateSpace Independent Publishing Platform.

Garrette, B., Phelps, C., & Sibony, O. (2018). *Cracked it!: How to solve big problems and sell solutions like top strategy consultants*. Palgrave Macmillan.

Kahneman, D. (2011). *Thinking, Fast and Slow*. Farrar, Straus and Giroux.

Kaufman, S.B. And Gregoire, C. (2016). *Wired to Create: Unraveling the Mysteries of the Creative Mind*. TarcherPerigee

Koch, R. (2005). *The 80/20 Individual: How to Build on the 20% of what You Do Best*. Crown Publishing Group.

Koch, R. (2011). *Living the 80/20 Way: Work Less, Worry Less, Succeed More, Enjoy More*. Quercus.

Koch, R. (2013). *The 80/20 Principle and 92 Other Powerful Laws of Nature: The Science of Success*. Quercus.

Lehrer, J. (2012). *Imagine. How Creativity Works*. Houghton Mifflin

Michalko, M. (2006). *Thinkertoys: A Handbook of Creative-Thinking Techniques (2nd Edition)*. Ten Speed Press.

Michalko, M. (2011). *Creative Thinkering: Putting Your Imagination to Work*. New World Library.

Nisbett, R. (2016). *Mindware Tools for Smart Thinking*. Farrar, Straus and Giroux.

Osborn, A. F. (1963). *Applied imagination: Principles and procedures of creative problem solving (Third Revised Edition)*. Charles Scribner's Sons.

Puccio, G. *The Creative Thinkers Toolkit*. Available through Great Courses at www.thegreatcourses.com/courses/the-creative-thinker-s-toolkit.html.

Saaty, T. L. (2012). *Decision Making for Leaders: The Analytic Hierarchy Process for Decisions in a Complex World*. RWS Publications.

Sawyer, K. (2013). *Zig Zag: The Surprising Path to Greater Creativity*. Jossey-Bass.

Silverstein, D., Samuel, P., & DeCarlo, N. (2012). *The Innovator's Toolkit: 50+ Techniques for Predictable and Sustainable Organic Growth*. Wiley.

Tauber, E. M. (1972). HIT: Heuristic Ideation Technique. A Systematic Procedure for New Product Search. *Journal of Marketing, 36*, 61-58.

Vogel, T. (2014). *Breakthrough Thinking: A Guide to Creative Thinking and Idea Generation*. HOW Books.

Von Oech, R. (2008). *A Whack on the Side of the Head: How You Can Be More Creative*. Grand Central Publishing.

ACKNOWLEDGEMENTS

THERE ARE MANY, many people who made this book possible. As a scientist, I owe a debt to my teachers at the University of Waterloo who started me on my career in biology, to my colleagues at the University of Guelph who empowered me to apply my knowledge, to all of my students who enriched my research and broadened my perspective, to my peers who supported, criticized, motivated and challenged me and to my colleagues at BASF in North Carolina, Germany and Belgium who worked with me as we explored the unknown together.

Jim Lewis, Lewis Institute, provided my training in project management in several courses offered by North Carolina State University. Doug DeCarlo helped me envision how project management might be applied to research. Together, Jim and Doug inspired me to write this book series.

I owe a special thanks to several people who painstakingly read drafts of this book. To Ed Kendall for his thorough reading and comments. To Suzanne Cunningham for encouragement, comments and lots of edits. To Joris de Wolf for all of the discussions about philosophy and statistics. To Doug DeCarlo for his suggestions on the text. To Joke Baute, Jan Chojecki, Tom Beeckman and Marijke Lein for their comments and suggestions on earlier versions.

ABOUT THE AUTHOR

BRYAN RECEIVED HIS B.Sc. and PhD degrees in biology from the University of Waterloo, Canada. He joined the faculty of the University of Guelph in the Crop Science Department in 1978, teaching graduate and undergraduate students and serving as Chairman of the Interdepartmental Plant Physiology and Director of the Plant Biotechnology Center during his 22 years in academic life. In 1999, he joined BASF Plant Science LLC in North Carolina USA as they started their research and development program in plant biotechnology. He worked as project coordinator for several international research programs with teams in Germany, Belgium and North Carolina. In 2007, he was a research manager in the Monsanto-BASF collaboration. Bryan has published over 100 peer-reviewed publications, numerous book chapters and 2 books, *Stress and Stress Coping in Cultivated Plants* written with Ya'acov Leshem (1994) and *Biotechnology and the Improvement of Forage Legumes* edited with DCW Brown (1997). He is an inventor on 11 issued US patents in plant biotechnology. Bryan retired in 2014 and lives in Jacksonville, Florida with his wife Marie. They have 3 children and 5 grandchildren, all living in Ontario, Canada.

www.ingramcontent.com/pod-product-compliance
Lightning Source LLC
Chambersburg PA
CBHW052312220526
45472CB00001B/86